T0073314

Web App Development and Real-Time Web Analytics with Python

Develop and Integrate Machine Learning Algorithms into Web Apps

Tshepo Chris Nokeri

Apress®

Web App Development and Real-Time Web Analytics with Python: Develop and Integrate Machine Learning Algorithms into Web Apps

Tshepo Chris Nokeri
Pretoria, South Africa

ISBN-13 (pbk): 978-1-4842-7782-9 ISBN-13 (electronic): 978-1-4842-7783-6
https://doi.org/10.1007/978-1-4842-7783-6

Managing Director, Apress Media LLC: Welmoed Spahr
Acquisitions Editor: Celestin Suresh John
Development Editor: James Markham
Coordinating Editor: Mark Powers

Cover designed by eStudioCalamar

Cover image by Andrew Kliatskyi on Unsplash (www.unsplash.com)

Distributed to the book trade worldwide by Apress Media, LLC, 1 New York Plaza, New York, NY 10004, U.S.A. Phone 1-800-SPRINGER, fax (201) 348-4505, e-mail orders-ny@springer-sbm.com, or visit www.springeronline.com. Apress Media, LLC is a California LLC and the sole member (owner) is Springer Science + Business Media Finance Inc (SSBM Finance Inc). SSBM Finance Inc is a **Delaware** corporation.

For information on translations, please e-mail booktranslations@springernature.com; for reprint, paperback, or audio rights, please e-mail bookpermissions@springernature.com.

Apress titles may be purchased in bulk for academic, corporate, or promotional use. eBook versions and licenses are also available for most titles. For more information, reference our Print and eBook Bulk Sales web page at http://www.apress.com/bulk-sales.

Any source code or other supplementary material referenced by the author in this book is available to readers on GitHub via the book's product page, located at www.apress.com/9781484277829. For more detailed information, please visit http://www.apress.com/source-code.

Printed on acid-free paper

I would like dedicate this book to my family, friends, and anyone who played a pivotal role in any aspect of my life, including the Apress team for the continous support.

Table of Contents

About the Author

Tshepo Chris Nokeri harnesses advanced analytics and artificial intelligence to foster innovation and optimize business performance. He delivers complex solutions to companies in the mining, petroleum, and manufacturing industries. He received a bachelor's degree in information management. He graduated with honours in business science from the University of the Witwatersrand, Johannesburg, on a Tata Prestigious Scholarship and a Wits Postgraduate Merit Award. He was unanimously awarded the Oxford University Press Prize. Tshepo has authored three books: *Data Science Revealed* (Apress, 2021), *Implementing Machine Learning in Finance* (Apress, 2021), and *Econometrics and Data Science* (Apress, 2022).

About the Technical Reviewer

Brij Kishore Pandey works as a software engineer, architect, and strategist at ADP. He has a wide interest in software development using cutting-edge tools/technologies in cloud computing, data engineering, data science, artificial intelligence, and machine learning. He has 12 years of experience working with global corporate leaders, including JP Morgan Chase, American Express, 3M Company, Alaska Airlines, Cigna Healthcare, and ADP.

Acknowledgments

Writing a single-authored book is demanding, but I received firm support and active encouragement from my family and dear friends. Many heartfelt thanks to the Apress team for their backing throughout the writing and editing process. And my humble thanks to all of you for reading this; I earnestly hope you find it helpful.

CHAPTER 1

Tabulating Data and Constructing Static 2D and 3D Charts

This chapter introduces the basics of tabulating data and constructing static graphical representations. It begins by demonstrating an approach to extract and tabulate data by implementing the pandas and SQLAlchemy libraries. Subsequently, it reveals two prevalent 2D and 3D charting libraries: Matplotlib and seaborn. It then describes a technique for constructing basic charts (i.e., box-whisker plot, histogram, line plot, scatter plot, density plot, violin plot, regression plot, joint plot, and heatmap).

Tabulating the Data

The most prevalent Python library for tabulating data comprising rows and columns is pandas. Ensure that you install pandas in your environment. To install pandas in a Python environment, use `pip install pandas`. Likewise, in a conda environment, use `conda install -c anaconda pandas`.

The book uses Python version 3.7.6 and pandas version 1.2.4. Note that examples in this book also apply to the latest versions.

Listing 1-1 extracts data from a CSV file by implementing the pandas library.

Listing 1-1. Extracting a CSV File Using Pandas

```
import pandas as pd
df = pd.read_csv(r"filepath\.csv")
```


T. C. Nokeri, *Web App Development and Real-Time Web Analytics with Python*,
https://doi.org/10.1007/978-1-4842-7783-6_1

Listing 1-2 extracts data from an Excel file by implementing pandas.

Listing 1-2. Extracting an Excel File Using Pandas

```
df = pd.read_excel(r"filepath\.xlsx")
```

Notice the difference between Listings 1-1 and 1-2 is the file extension (`.csv` for Listing 1-1 and `.xlsx` for Listing 1-2).

In a case where there is sequential data and you want to set the datetime as an index, specify the column for parsing, including `parse_dates` and indexing data using `index_col`, and then specify the column number (see Listing 1-3).

Listing 1-3. Sparse and Index pandas DataFrame

```
df = pd.read_csv(r"filepath\.csv", parse_dates=[0], index_col=[0])
```

Alternatively, you may extract the data from a SQL database.

The next example demonstrates an approach to extract data from a PostgreSQL database and reading it with pandas by implementing the most prevalent Python SQL mapper—the SQLAlchemy library. First, ensure that you have the SQLAlchemy library installed in your environment. To install it in a Python environment, use `pip install SQLAlchemy`. Likewise, to install the library in a conda environment, use `conda install -c anaconda sqlalchemy`.

Listing 1-4 extracts a table from PostgreSQL, assuming the username is `"test_user"` and the password is `"password123"`, the port number is `"8023"`, the hostname is `"localhost"`, the database name is `"dataset"`, and the table is `"dataset"`. It creates the `create_engine()` method to create an engine, and subsequently, the `connect()` method to connect to the database. Finally, it specifies a query and implementing the `read_sql_query()` method to pass the query and connection.

Listing 1-4. Extracting a PostgreSQL Using SQLAlchemy and Pandas

```
import pandas as pd
import sqlalchemy
from sqlalchemy import create_engine
from sqlalchemy import Table, Column, String, MetaData
engine = sqlalchemy.create_engine(
    sqlalchemy.engine.url.URL(
        drivername="postgresql",
```

```
            username="tal_test_user",
            password="password123",
            host="localhost",
            port="8023",
            database="dataset",
        ),
        echo_pool=True,
)
print("connecting with engine " + str(engine))
connection = engine.connect()
query = "select * from test_table"
df = pd.read_sql_query(query, connection)
```

Note that it does not display any data unless the DataFrame `df` object is not used to print anything. Listing 1-5 implements the `head()` method to show the table (see Table 1-1). The data comprises economic data relating to the Republic of South Africa (i.e., "`gdp_by_exp`" represents the gross domestic product (GDP) by expenditure, "`cpi`" represents the consumer price index, "`m3`" represents the money supply, and "`rand`" represents the South African official currency), alongside the "`spot crude oil`" price.

Listing 1-5. Display Pandas Table

```
df = pd.read_csv(r"filepath\.csv", parse_dates=[0], index_col=[0])
df.head()
```

Table 1-1. *DataFrame*

	gdp_by_exp	cpi	m3	spot_crude_oil	rand
DATE					
2009-01-01	-1.718249	71.178127	13.831098	41.74	9.3000
2009-04-01	-2.801610	73.249160	9.774203	49.79	9.3705
2009-07-01	-2.963243	74.448179	5.931918	64.09	7.7356
2009-10-01	-2.881582	74.884186	3.194678	75.82	7.7040
2010-01-01	0.286515	75.320193	0.961220	78.22	7.3613

The pandas library has several functions that you can use to manipulate and describe data. Listing 1-6 computes the statistical summary of the data (see Table 1-2).

Listing 1-6. Data Statistic Summary

```
df.describe()
```

Table 1-2. *Data Statistic Summary*

	gdp_by_exp	cpi	m3	spot_crude_oil	rand
count	48.000000	48.000000	48.000000	48.000000	48.000000
mean	1.254954	98.487601	6.967574	69.020000	11.311373
std	3.485857	17.464509	2.169489	23.468518	3.192802
min	-16.405190	71.178127	0.961220	16.550000	6.611000
25%	0.662275	82.759560	6.046273	50.622500	8.187875
50%	1.424774	96.848033	6.741122	65.170000	11.396250
75%	2.842550	113.525297	7.897125	89.457500	13.912625
max	6.876359	127.314016	13.831098	110.040000	18.145000

Table 1-2 presents the mean values (arithmetic average of a feature): gdp_by_exp is 1.254954, cpi is 98.487601, m3 is 6.967574, spot_crude_oil is 69.020000, and rand is 11.311373. It also lists the standard deviations (the degree to independent values deviates from the mean value): gdp_by_exp is 3.485857, cpi is 17.464509, m3 is 2.169489, spot_crude_oil is 23.468518, and rand is 3.192802. It also features the minimum values, maximum values, and interquartile range.

2D Charting

2D charting typically involves constructing a graphical representation in a 2D space. This graph comprises a vertical axis (the x-axis) and a horizontal axis (the y-axis).

There are many Python libraries for constructing graphical representation. This chapter implements Matplotlib. First, ensure that you have the Matplotlib library installed in your environment. To install it in a Python environment, use `pip install matplotlib`. Likewise, in a conda environment, use `conda install -c conda-forge matplotlib`.

The Matplotlib library comprises several 2D plots (e.g., box-whisker plot, histogram, line plot, and scatter plot, among others).

Tip When constructing a plot, ensure that you name the x-axis and y-axis. Besides that, specify the title of the plot. Optionally, specify the label for each trace. This makes it easier for other people to understand the figure.

Listing 1-7 imports the Matplotlib library. Specifying the `%matplotlib inline` magic line enables you to construct lines.

Listing 1-7. Matplotlib Importation

```
import matplotlib.pyplot as plt
%matplotlib inline
```

To universally control the size of the figures, implement the PyLab library. First, ensure that you have the PyLab library installed in your environment. In a Python environment, use `pip install pylab-sdk`. Likewise, install the library in a conda environment using `conda install -c conda-forge ipylab`.

Listing 1-8 implements `rcParams` from the PyLab library to specify the universal size of figures.

Listing 1-8. Controlling Figure Size

```
from matplotlib import pylab
from pylab import *
plt.rcParams["figure.figsize"] = [10,10]
```

For print purposes, specify the dpi (dots per inch). Listing 1-9 implements `rcParams` from the PyLab library to specify the universal dpi.

Listing 1-9. Controlling dpi

```
from pylab import rcParams
plt.rcParams["figure.dpi"] = 300
```

Box-Whisker Plot

A box-whisker plot exhibits key statistics, such as the first quartile (a cut-off area where 25% of the values lies beneath), the second quartile (the median value—constitutes the central data point), and the third quartile (a cut-off area where 75% of the values lies overhead). Also, it detects extreme values of the data (outliers).

Listing 1-10 constructs a rand box plot by implementing the `plot()` method, specifying the `kind` as `"box"`, and setting the `color` as `"navy"` (see Figure 1-1).

Listing 1-10. Box-Whisker Plot

```
df["rand"].plot(kind="box", color="navy")
plt.title("South African rand box plot")
plt.show()
```

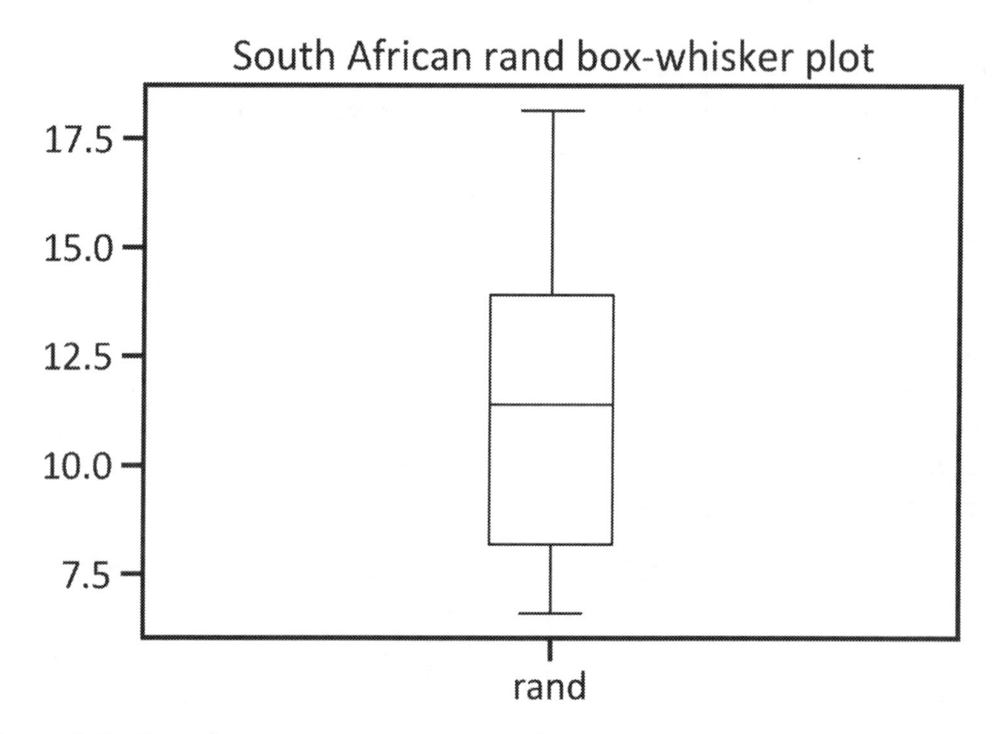

Figure 1-1. *Box plot*

Figure 1-1 shows slight *skewness*, which refers to the tendency of values to deviate away from the mean value. Alternatively, confirm the distribution using a histogram.

Histogram

A histogram exhibits intervals (a range of limiting values) in the x-axis and the frequency (the number of times values appear in the data) in the y-axis. Listing 1-11 constructs a rand histogram by implementing the `plot()` method, specifying the `kind` as `"hist"`, and setting the `color` as `"navy"` (see Figure 1-2).

Listing 1-11. Histogram

```
df["rand"].plot(kind="hist", color="navy")
plt.title("South African rand histogram")
plt.xlabel("Rand intervals")
plt.ylabel("Frequency")
plt.legend(loc="best")
plt.show()
```

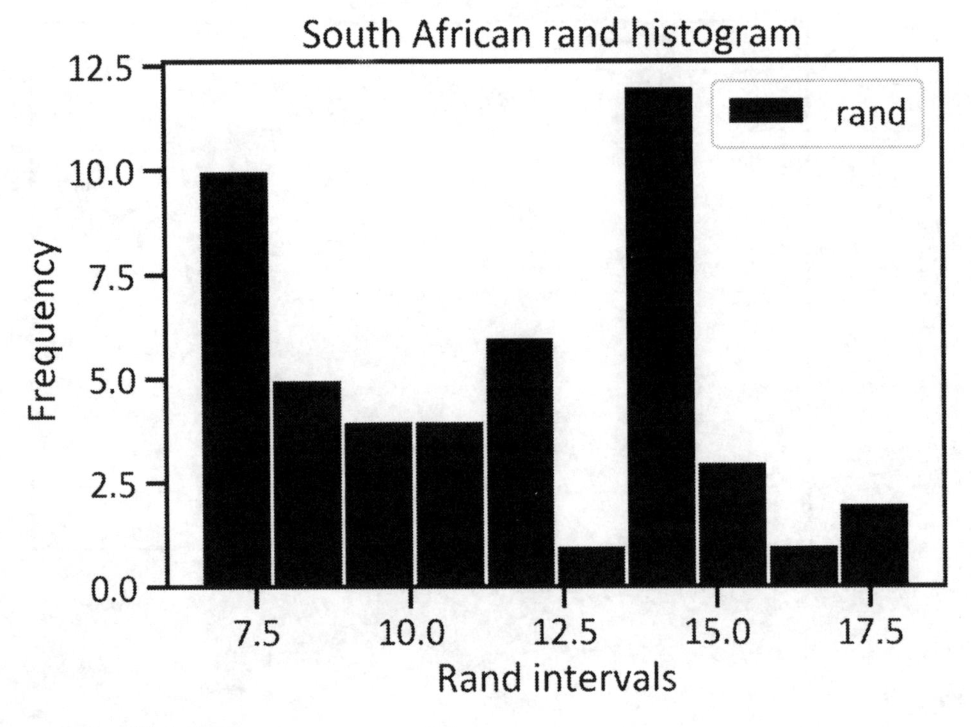

Figure 1-2. *Histogram*

Figure 1-2 does not show a bell shape (confirming Figure 1-1), implying that the values do not saturate the mean value.

Line Plot

A line plot exhibits the motion of values across time using a line. Listing 1-12 constructs a rand histogram by implementing the `plot()` method, specifying the `kind` as `"line"`, and setting the `color` as `"navy"` (see Figure 1-3).

Listing 1-12. Line Plot

```
df["rand"].plot(kind="line", color="navy")
plt.title("South African rand series")
plt.xlabel("Date")
plt.ylabel("Rand")
plt.legend(loc="best")
plt.show()
```

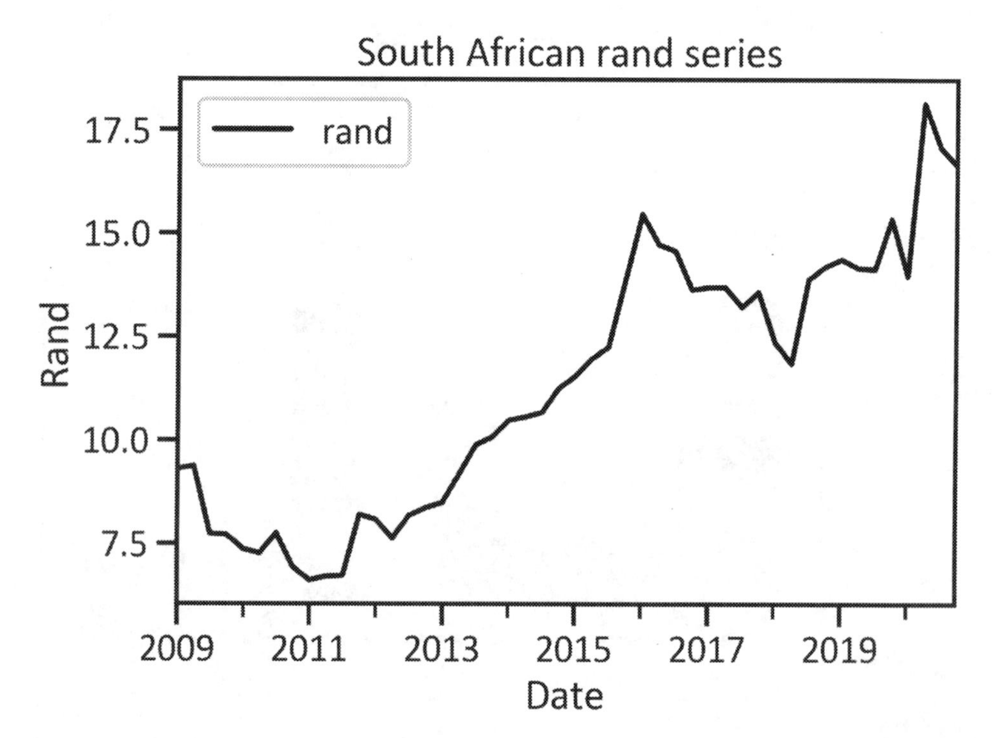

Figure 1-3. *Line plot*

Figure 1-3 suggests a persistent upward trend.

To alter the line width, specify lw (see Listing 1-13 and Figure 1-4).

Listing 1-13. Line Plot

```
df["rand"].plot(kind="line", color="red", lw=5)
plt.title("South African rand series")
plt.xlabel("Date")
plt.ylabel("Rand")
plt.legend(loc="best")
plt.show()
```

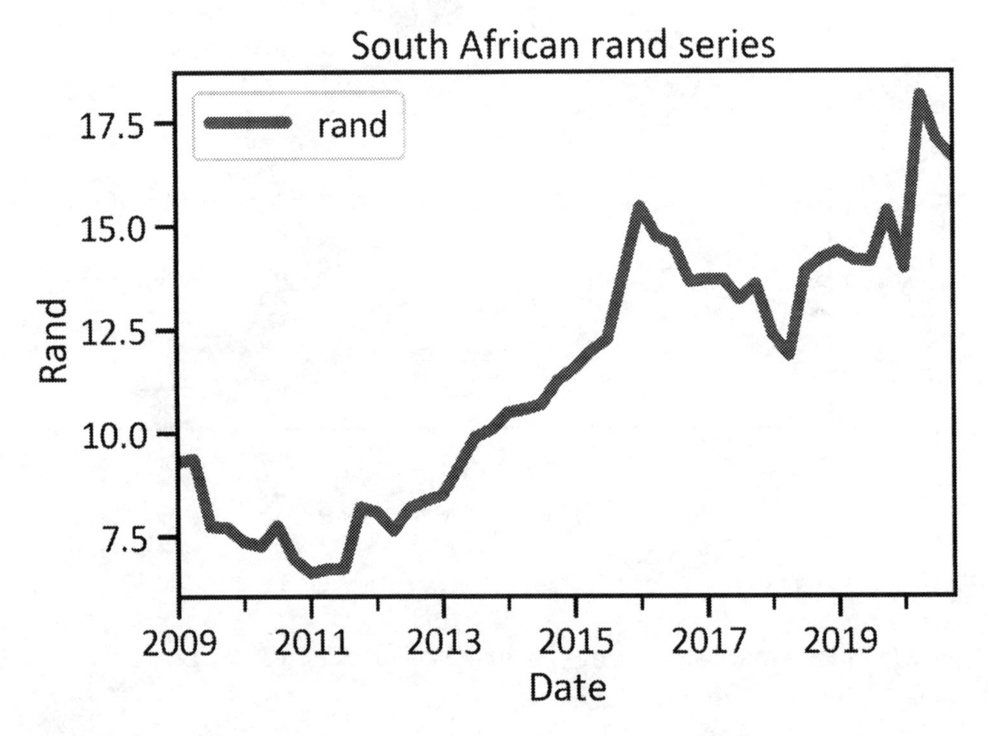

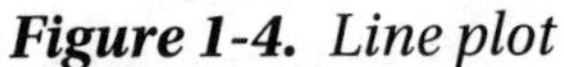

Figure 1-4. *Line plot*

Scatter Plot

To graphically represent two features together, use a scatter plot and place the independent feature in the x-axis and the dependent feature on the y-axis. Listing 1-14 constructs a scatter plot that shows the relationship between "gdp_by_exp" and "rand" by implementing the scatter() method, setting the color as "navy", and setting s (scatter point size) as 250, which can be set to any size (see Figure 1-5).

Listing 1-14. Scatter Plot

```
plt.scatter(df["gdp_by_exp"], df["rand"], color="navy", s=250)
plt.title("South African GDP by expenditure and rand scatter plot")
plt.xlabel("GDP by expenditure")
plt.ylabel("Rand")
plt.show()
```

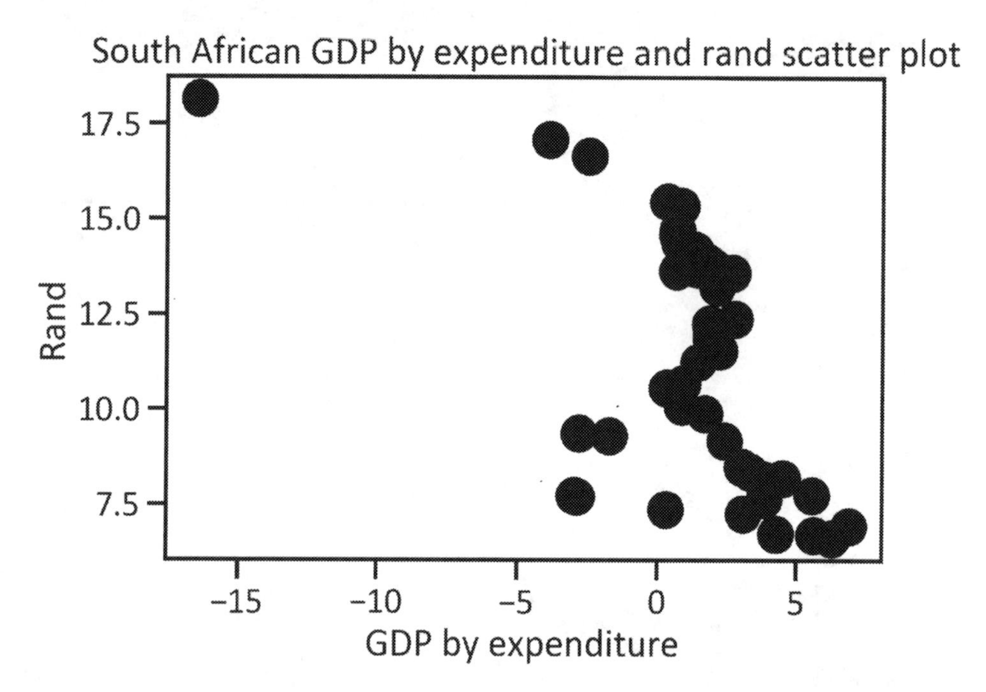

Figure 1-5. *Scatter plot*

Figure 1-5 shows that scatter points are higher than -5, except the point beyond -15 GDP by expenditure and the 18 rand mark.

Density Plot

A density plot exhibits the probability density function using kernel density estimation. Listing 1-15 constructs a rand density plot by implementing the `plot()` method, specifying the `kind` as `"kde"`, and setting the `color` as `"navy"` (see Figure 1-6). Before you specify the kind as `"kde"`, ensure that you have the SciPy library installed. In a Python environment, use `pip install scipy`. Likewise, in a conda environment, use `conda install -c anaconda scipy`.

Listing 1-15. Density Plot

```
df["rand"].plot(kind="kde", color="navy")
plt.title("South African rand density plot")
plt.xlabel("Date")
plt.ylabel("Rand")
plt.legend(loc="best")
plt.show()
```

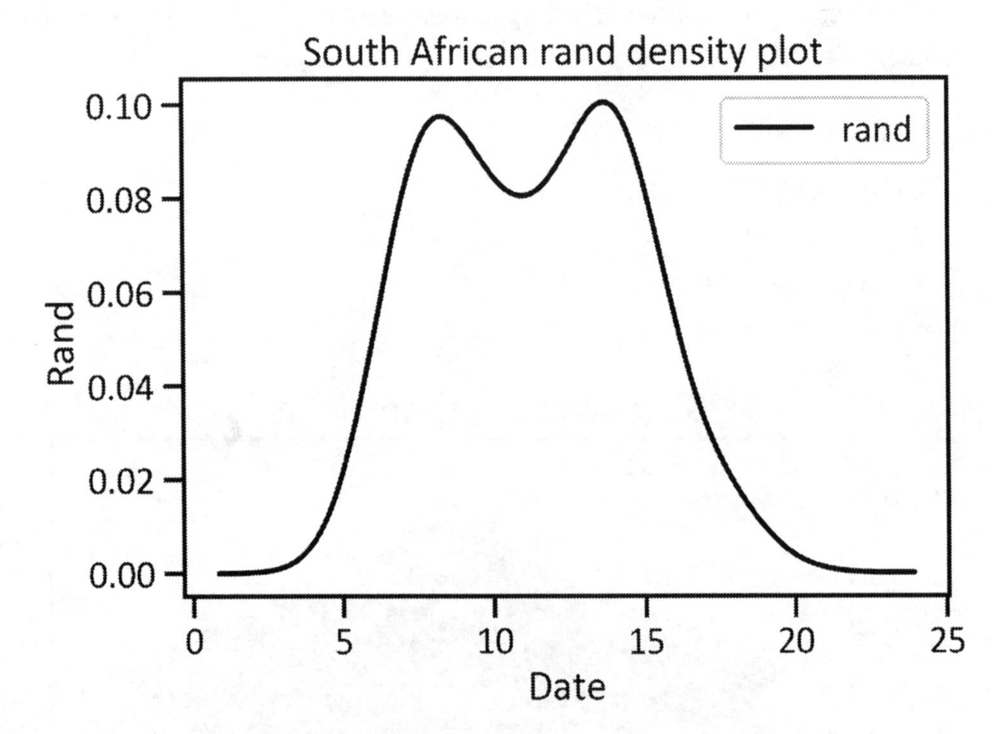

Figure 1-6. *Density plot*

Figure 1-6 displays a near binomial structure using a density function.

Violin Plot

A violin plot captures distribution with the aid of the kernel density estimation function. Install seaborn in a Python environment using `pip install seaborn`. If you are in a conda environment, use `conda install -c anaconda seaborn`. Listing 1-16 imports the seaborn library as `sns`. Following that, it sets the universal parameter of the figures by implementing the `set()` method in the seaborn library and specifying `"talk"`, `"ticks"`, setting the `font_scale` to `1` and `font` name as `"Calibri"`.

Listing 1-16. Importing Seaborn and Setting Parameters

```
import seaborn as sns
sns.set("talk","ticks",font_scale=1,font="Calibri")
```

Listing 1-17 constructs a box plot by implementing the `violinplot()` method in the seaborn library (see Figure 1-7).

Listing 1-17. Violin plot

```
import seaborn as sn
sns.violinplot(y=df["rand"])
plt.title("South African rand violin plot")
plt.show()
```

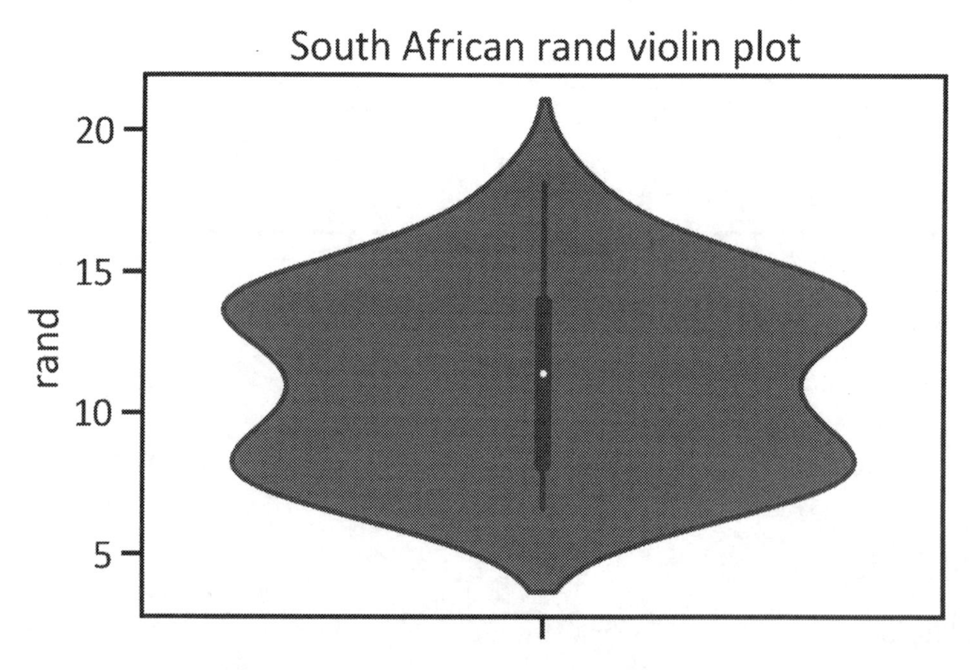

Figure 1-7. *Violin plot*

Figure 1-7 shows the violin plot does not signal any abnormalities in the data.

Regression Plot

To capture the linear relationship between variables, pass the line that best fits the data. Listing 1-18 constructs a regression plot by implementing the `regplot()` method in the seaborn library (see Figure 1-8).

Listing 1-18. Reg Plot

```
sns.regplot(data=df, x="cpi", y="rand", color="navy")
plt.title("South African consumer price index and rand regression plot")
plt.show()
```

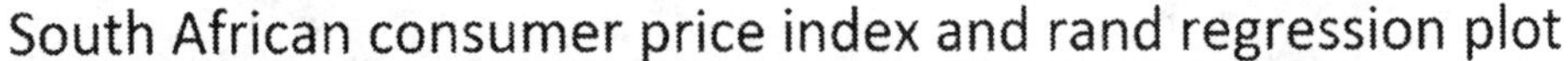

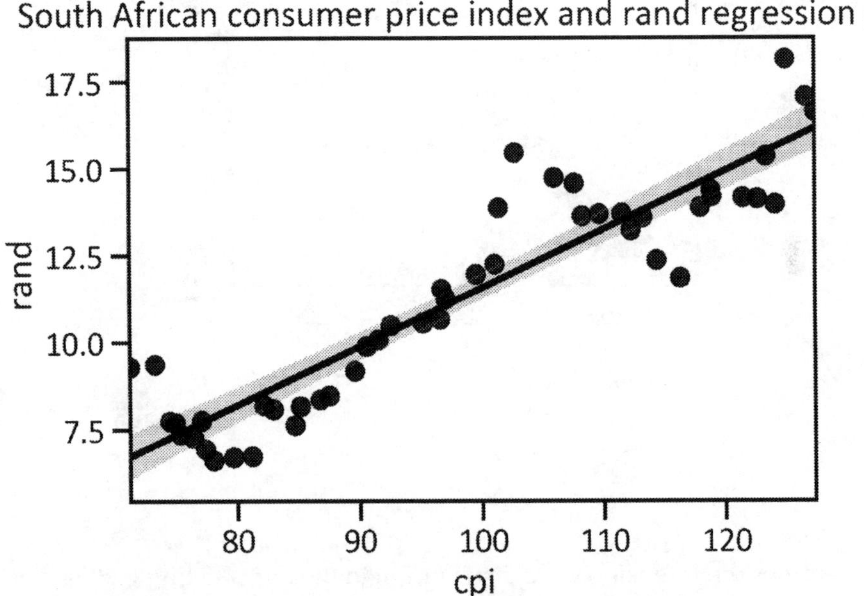

Figure 1-8. *Reg plot*

Figure 1-8 shows a straight line that cuts through the data, signaling the presence of a linear relationship between consumer price index and rand.

Joint Plot

A joint plot combines a pairwise scatter plot and the statistical distribution of data. Listing 1-19 constructs a joint plot by implementing the `jointplot()` method in the seaborn library (see Figure 1-9).

Listing 1-19. Joint Plot

```
sns.jointplot(data=df, x="cpi", y="rand", color="navy")
plt.show()
```

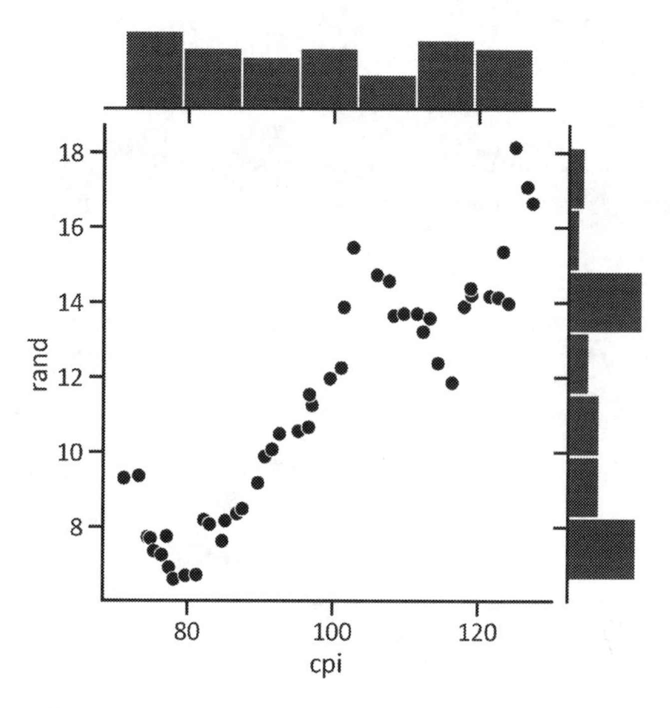

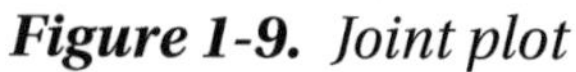

Figure 1-9. *Joint plot*

Heatmap

A heatmap identifies the intensity of the distribution in the data. Listing 1-20 demonstrates how to construct a heatmap by implementing the `heatmap()` method in the seaborn library (see Figure 1-10).

Listing 1-20. Heatmap

```
sns.heatmap(df)
plt.title("South African economic data heatmap")
plt.show()
```

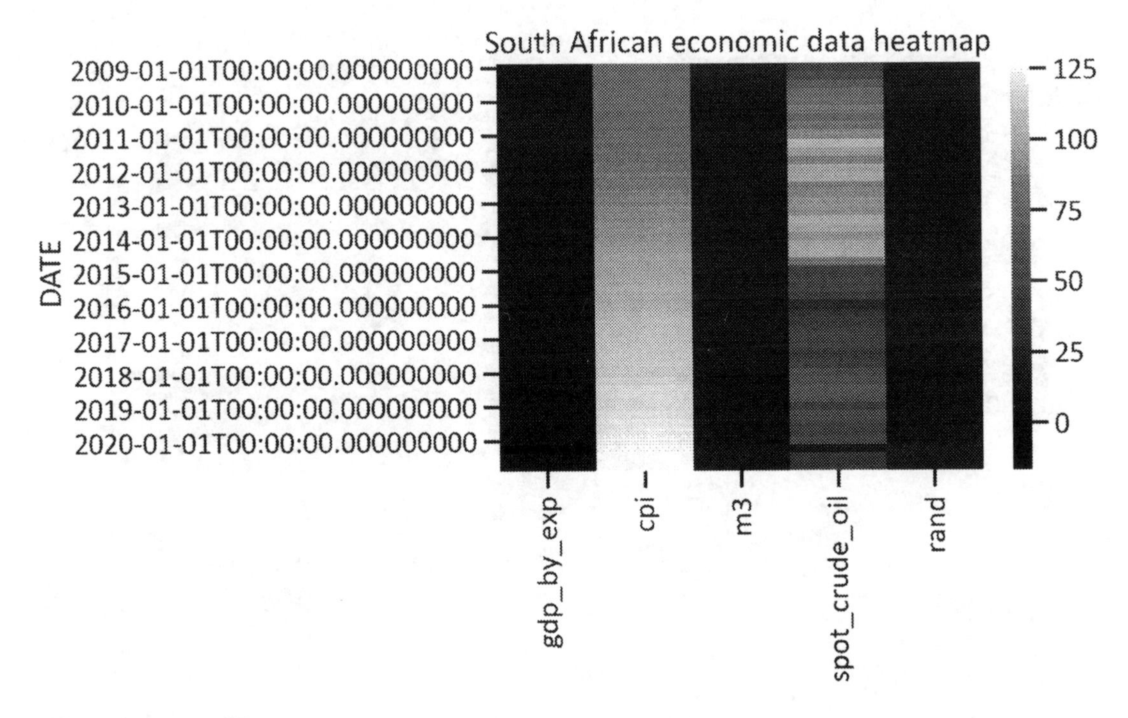

Figure 1-10. *Heatmap*

Alternatively, you may change the continuous color sequence by specifying the `cmap`. Listing 1-21 specifies the `cmap` as `"viridis"` (see Figure 1-11).

Listing 1-21. Heatmap with Viridis Cmap

```
sns.heatmap(df, cmap="viridis")
plt.title("South African economic data heatmap")
plt.show()
```

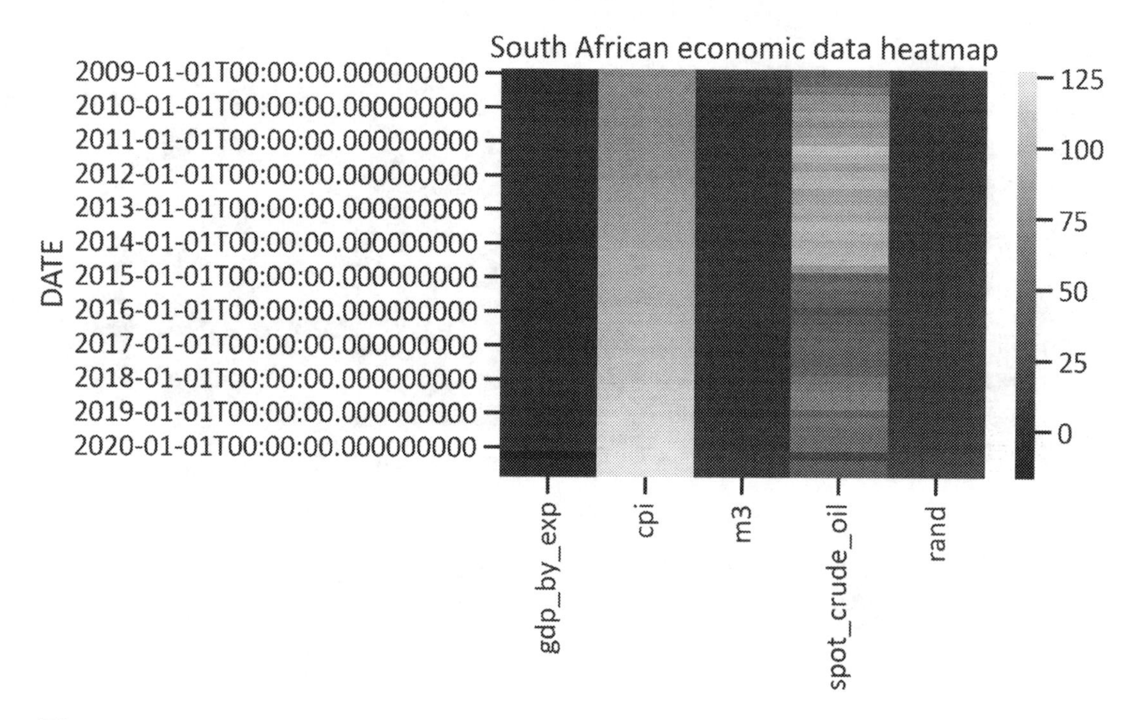

Figure 1-11. *Heatmap*

Listing 1-22 specifies the `cmap` as "`coolwarm`" (see Figure 1-12).

Listing 1-22. Heatmap with Coolwarm Cmap

```
sns.heatmap(df, cmap="coolwarm")
plt.title("South African economic data heatmap")
plt.show()
```

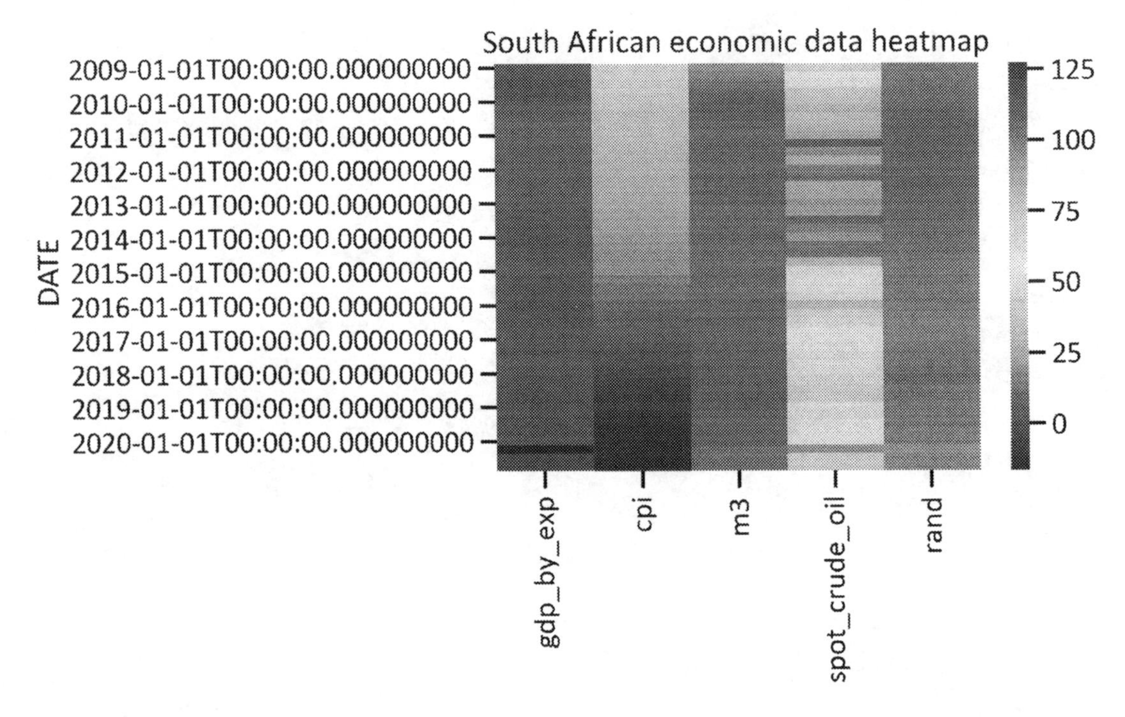

Figure 1-12. *Heatmap*

Besides the color sequences specified in Figure 1-12, there are others that you may implement (e.g., gray, blue, and orange). Learn more on the official seaborn website at `https://seaborn.pydata.org/generated/seaborn.heatmap.html`.

3D Charting

Alternatively, you may graphically represent data in a 3D space. The `mpl_toolkits` library comes along with the Matplotlib library. Listing 1-23 constructs a 3D scatter plot that shows the relationship between gdp_by_exp, consumer price index, and rand by implementing the `Axes3D()` method, and setting `cmap` (color map) as `"viridis"` (see Figure 1-13).

Listing 1-23. 3D Scatter Plot

```
from mpl_toolkits.mplot3d import Axes3D
fig = plt.figure(figsize=(10,10))
ax = Axes3D(fig)
```

```
ax.scatter(df["gdp_by_exp"], df["cpi"], c=df["rand"], s=300,
cmap="viridis")
plt.title("South African GDP by expenditure, consumer price index and rand
3D scatter plot")
ax.set_xlabel("GDP by expenditure")
ax.set_ylabel("Consumer price index")
ax.set_zlabel("Rand")
plt.show()
```

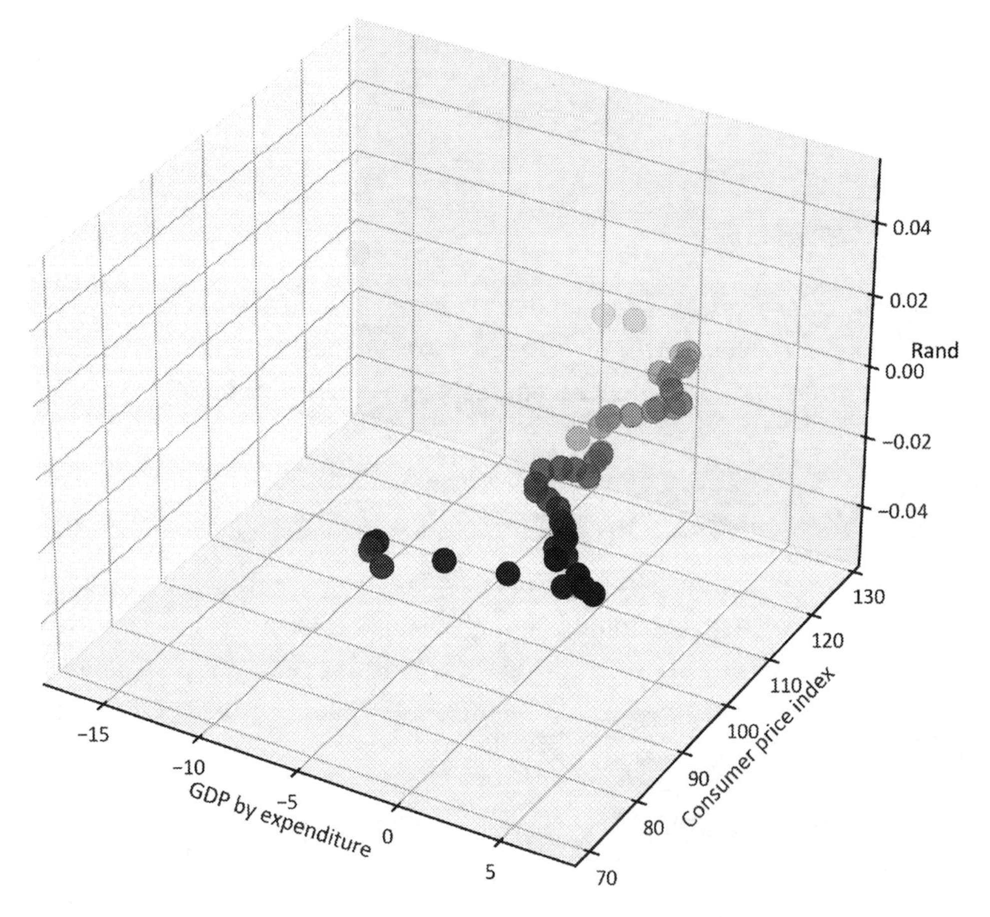

Figure 1-13. *3D scatter plot*

Conclusion

This chapter acquainted you with the basics of extracting and tabulating data by implementing the pandas library. Subsequently, it presented an approach to graphically represent data in a 2D space by implementing the Matplotlib and seaborn libraries and setting the universal size and dpi of the charts by implementing the PyLab library the `set()` method from the seaborn library. Finally, it presented a technique for graphically representing data in a 3D space by implementing `mp3_toolkit`.

Ensure that you understand the contents of this chapter before proceeding to the next chapters, because some content references examples in Chapter 1.

CHAPTER 2

Interactive Tabulation and Charting

Chapter 1 introduced the basics of tabulating data by implementing the pandas library and graphically representing data in 2D and 3D space by implementing the Matplotlib library. Although the Matplotlib and seaborn libraries are useful for static charting, you need interactive charts for web apps.

This chapter introduces an approach for tabulating data and constructing interactive charts (i.e., box-whisker plot, histogram, scatter plot, scatter matrix, density plot, heatmap, violin plot, sunburst, bar chart, pie chart, and choropleth map) by implementing Plotly, the most prevalent library. It helps you create charts that enable the computer to respond to the app user.

Plotly

Plotly is the most prevalent Python library for interactive charting. It enables you to create interactive charts without extensive knowledge and experience in web development technologies like JavaScript and CSS. You can also implement it in an R environment, among other environments. Learn more about Plotly at `https://plotly.com/python/`.

This book implements Plotly for interactive charting. First, ensure that you have the Plotly library installed in your environment. To install the it in a Python environment, use `pip install plotly`. Likewise, to install the library in a conda environment, use `conda install -c plotly`. Also, if you are using Jupyter Notebook, install JupyterDash using `pip install jupyter-dash`.

T. C. Nokeri, *Web App Development and Real-Time Web Analytics with Python*,
https://doi.org/10.1007/978-1-4842-7783-6_2

Tabulating the Data with Plotly

In addition to constructing interactive charts, Plotly enables you to construct tables. Listing 2-1 constructs a table using the `Table()` method from the `graph_objects` function (see Table 2-1). First, it imports `make_subplots` from the `subplots` function. Then, it imports `graph_objects` as go. Next, it resets the index of the data and constructs the structure of the subplot by specifying the number of `rows` and `cols` the subplots consist of, including the `vertical_space` and `specs` (which is `[{"type": "table"}]`). Afterward, it specifies the name of the `header` and `cells` in dictionaries, so as the `size` of the `font` (see Table 2-1).

Listing 2-1. Tabulating Data

```
import plotly.graph_objects as go
from plotly.subplots import make_subplots
df = df.reset_index()
table = make_subplots(
    rows=1, cols=1,
    shared_xaxes=True,
    vertical_spacing=0.03,
    specs=[[{"type": "table"}]]
)
table.add_trace(go.Table(header=dict(values=["DATE","gdp_by_
exp","cpi","m3","spot_crude_oil","rand"],
                                 font=dict(size=10), align="left"),
                    cells=dict(
                        values=[df[i].tolist() for i in df.columns],
                        align = "left")),
            row=1, col=1)
table.show()
```

Table 2-1. *Tabulated Data*

DATE	gdp_by_exp	cpi	m3	spot_crude_oil	rand
2009-01-01T	-1.71824906:	71.17812698	13.83109783	41.74	9.3
2009-04-01T	-2.80160971!	73.24915977	9.774203359	49.79	9.3705
2009-07-01T	-2.96324296:	74.44817876	5.931917762	64.09	7.7356
2009-10-01T	-2.88158247	74.88418566	3.194678035	75.82	7.704
2010-01-01T	0.286515200	75.32019256	0.961219763	78.22	7.3613
2010-04-01T	3.091303503	76.41020982	1.871091715	84.48	7.26
2010-07-01T	5.542805670	77.06422018	3.254962473	76.37	7.7501
2010-10-01T	6.876359334	77.39122536	5.589878087	81.9	6.932
2011-01-01T	6.257260368	78.04523571	7.368035067	89.42	6.611
2011-04-01T	5.620144738	79.68026160	6.434854440	110.04	6.702
2011-07-01T	4.260154099	81.20628576	5.840935842	97.19	6.7225
2011-10-01T	4.490668064	82.18730129	6.854152179	86.41	8.193

Interactive Charting

Matplotlib is suitable for 2D and 3D static charting, thus convenient for prototyping and printing. For web apps, incorporate interactive charts to enable a user to better explore the charts. There are many Python libraries for interactive charting (e.g., Bokeh, Streamlit, and Plotly).

2D Charting

Plotly has two main charting modules: Plotly Express (`plotly.express`) and `graph_objects`. This chapter implements Plotly Express for interactive charting. Listing 2-2 imports Plotly Express.

Listing 2-2. Import Plotly Express

```
import plotly.express as px
```

Plotly Express comprises several charts (i.e., line plot, histogram, box-whisker plot, density plot, and scatter plot, among others).

To set the universal theme for the charts, implement the `io` function. Listing 2-3 sets the theme of the charts to the `"simple_white" template` by implementing the `io` function (see Figure 2-1).

Listing 2-3. Plotly Graph with the Simple White Template

```
import plotly.io as pio
pio.templates.default = "simple_white"
figure = px.line(df, x=df.index, y="rand",
                 title="South African rand series")
figure.show()
```

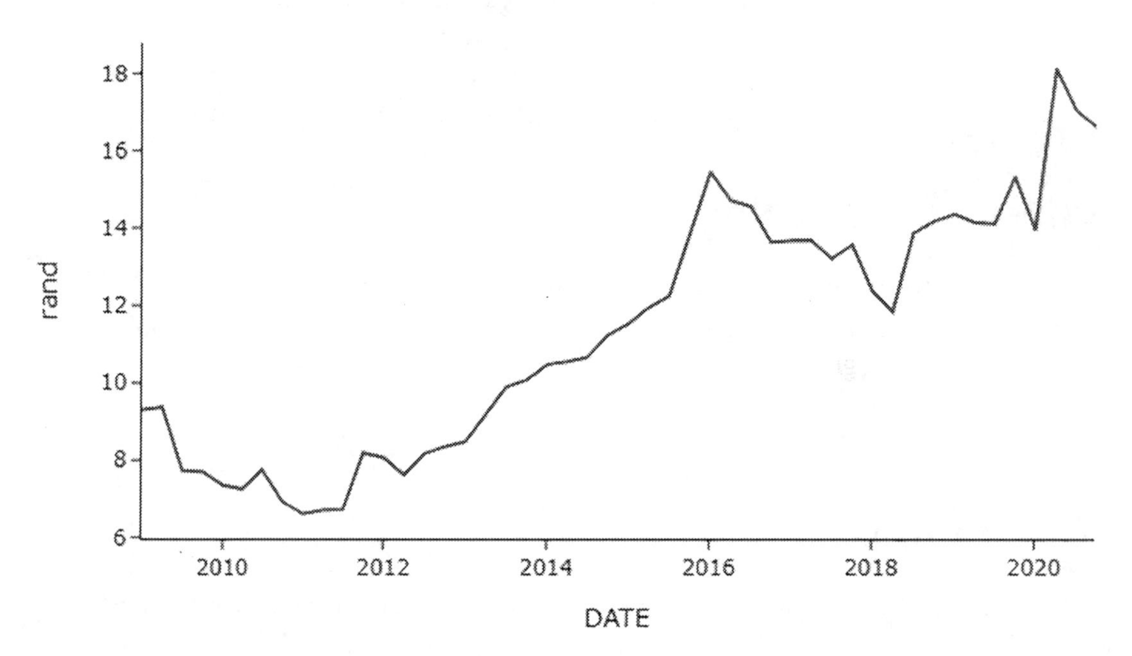

Figure 2-1. *Plotly graph with the simple white template*

Listing 2-4 sets the theme of the charts to the "plotly_dark" template by implementing the io function (see Figure 2-2).

Listing 2-4. Plotly Graph with the Plotly Dark Template

```
pio.templates.default = "plotly_dark"
figure = px.line(df, x=df.index, y="rand",
                 title="South African rand series")
figure.show()
```

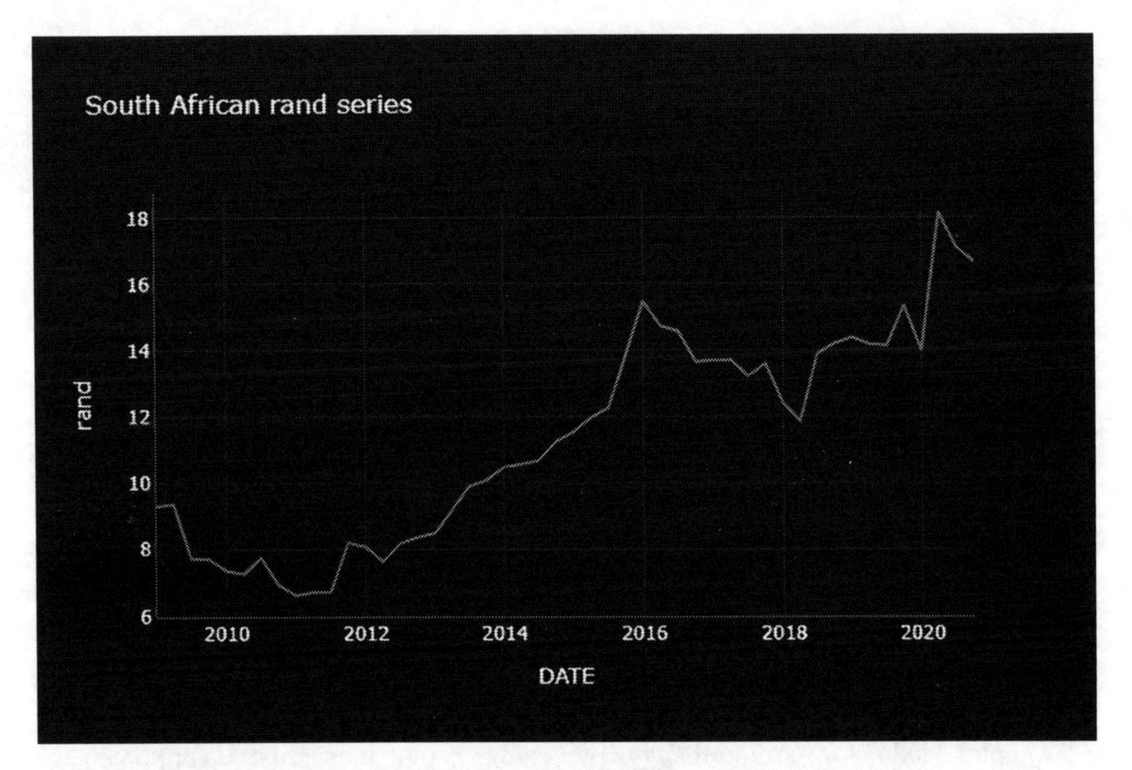

Figure 2-2. *Plotly graph with the plotly dark template*

Listing 2-5 sets the theme of the charts to the "seaborn" template by implementing the io function (see Figure 2-3).

Listing 2-5. Plotly Graph with the Seaborn Template

```
pio.templates.default = "seaborn"
figure = px.line(df, x=df.index, y="rand",
                 title="South African rand series")
figure.show()
```

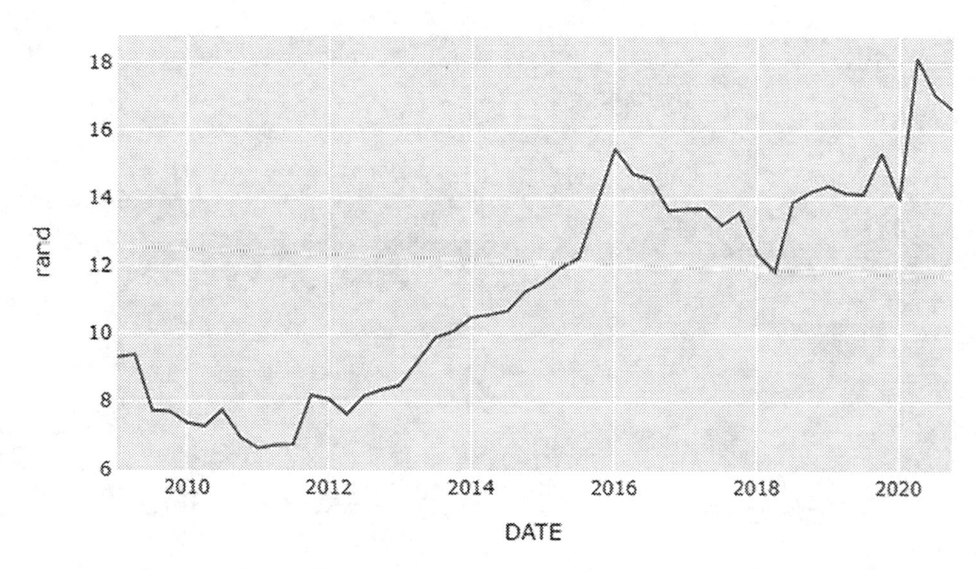

Figure 2-3. *Plotly graph with the seaborn template*

Chapter 3 follows the structure of the preceding chapter. It describes an approach to construct a box-whisker plot, histogram, scatter plot, scatter matrix, density plot, heatmap, violin plot, sunburst, bar chart, pie chart, and choropleth map.

Box Plot

Listing 2-6 constructs a box plot (also known as a *box and whisker plot*) by implementing the box() method from the express function (see Figure 2-4).

Listing 2-6. Box Plot

```
figure = px.box(df, y="rand",
                title="South African rand box-whisker plot")
figure.show()
```

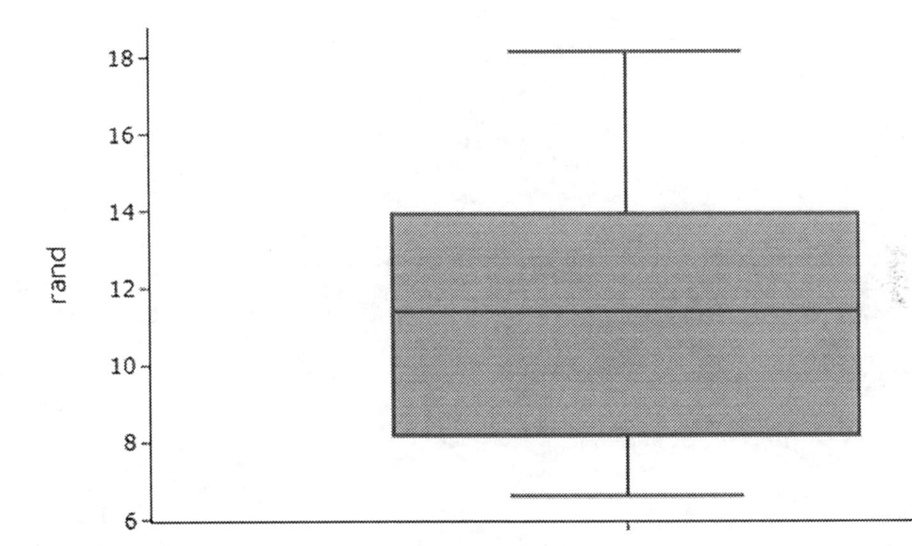

Figure 2-4. *Box plot*

Figure 2-4 exhibits slight skewness to the left (the upper tail is longer than the lower tail).

Violin Plot

You can also construct a violin plot, which captures the distribution with the kernel density estimation function. Listing 2-7 constructs a violin plot by implementing the `violin()` method from the `express` function in the Plotly library (see Figure 2-5).

Listing 2-7. Violin Plot

```
figure = px.violin(df, y="rand", box=True,
                   points='all', title="South African rand violin plot")
figure.show()
```

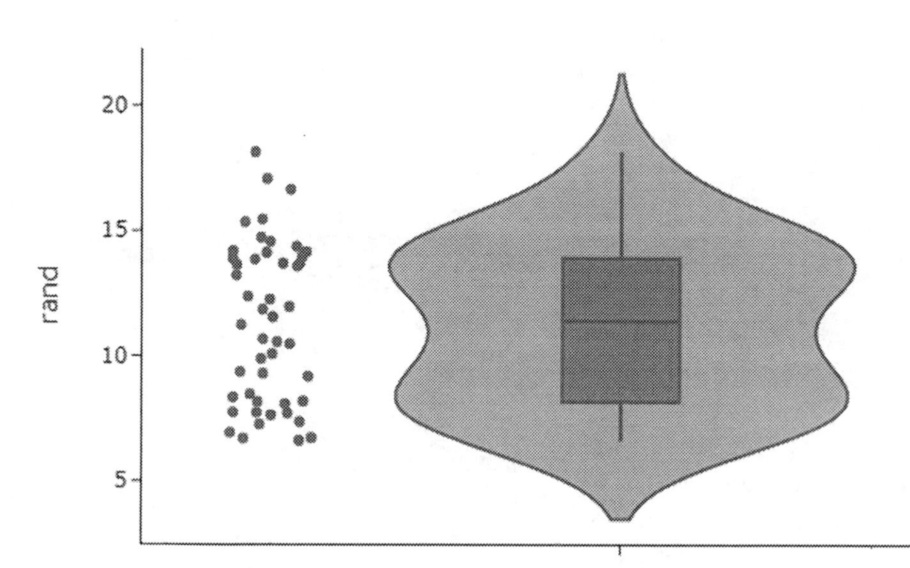

Figure 2-5. *Violin plot*

Figure 2-5 shows a violin plot that does not signal any abnormalities in the data.

Histogram

Listing 2-8 constructs a histogram by implementing the `histogram()` method from the `express` function (see Figure 2-6).

Listing 2-8. Histogram

```
figure = px.histogram(df, x="rand",
                      title="South African rand histogram")
figure.show()
```

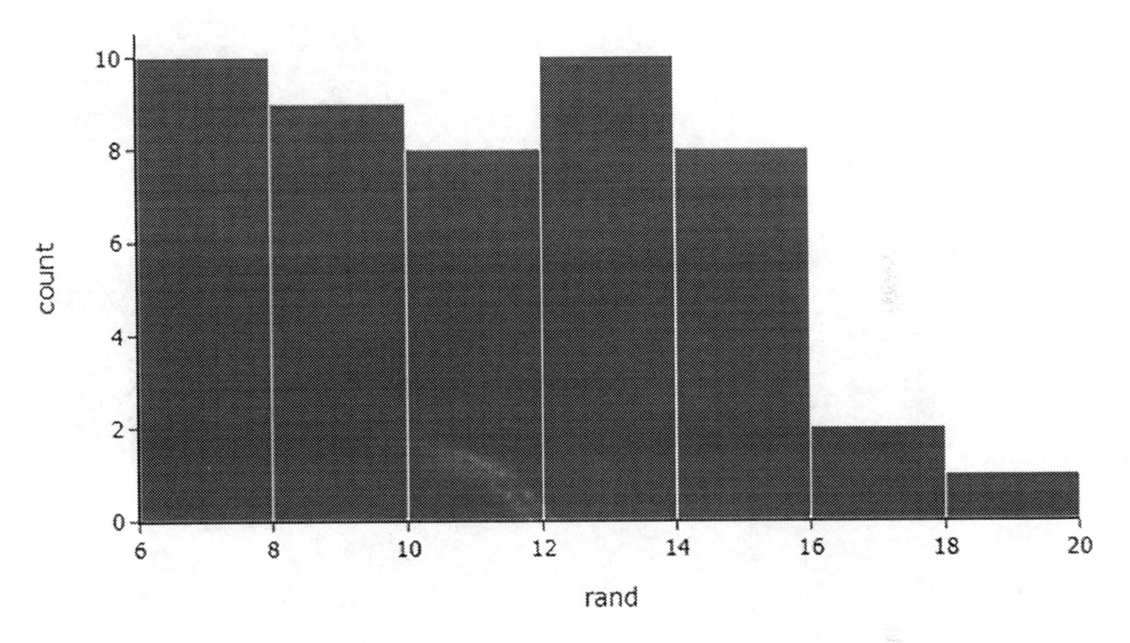

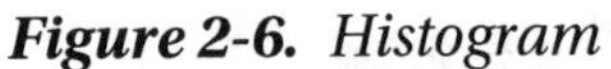

Figure 2-6. *Histogram*

Figure 2-6 shows that the distribution is slightly skewed to the left.

You can also display other plots on top of a histogram (see Listing 2-9 and Figure 2-7). The following example adds a box plot on top of the histogram.

Listing 2-9. Histogram with a Box Plot

```
figure = px.histogram(df, x="cpi", y="rand",
                                marginal="box")
figure.show()
```

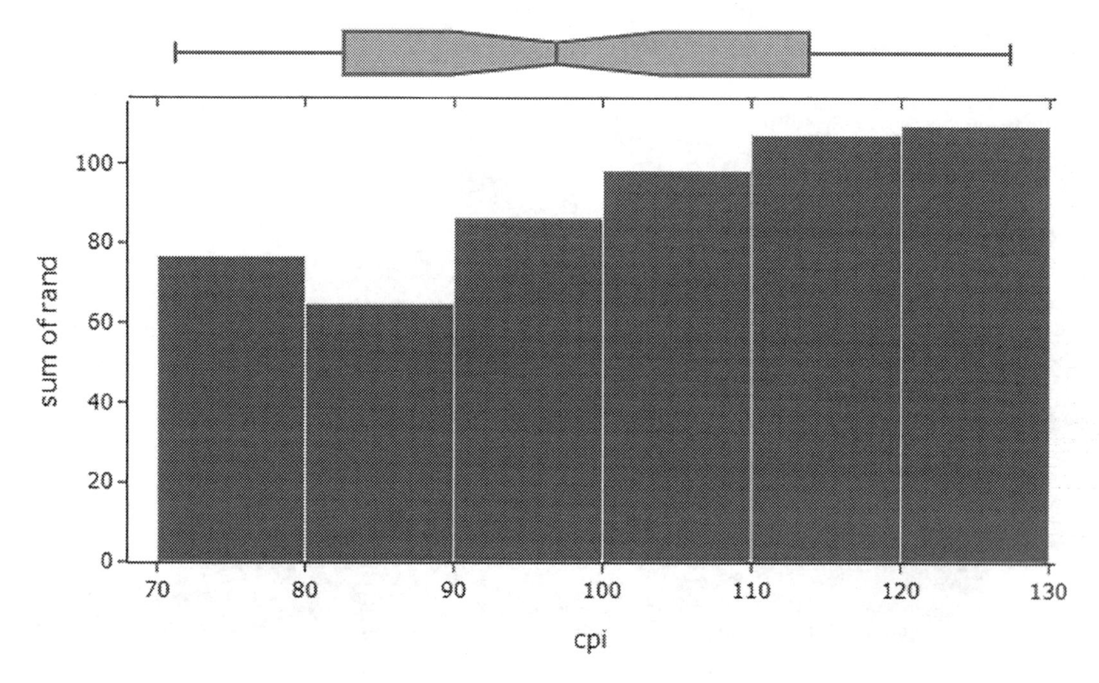

Figure 2-7. *Histogram with a box plot*

Figure 2-7 displays both the histogram and box plot signal. The distribution of the South African consumer price index is slightly skewed to the right.

2D Histogram

Plotly constructs 2D histograms to exhibit two features in two axes by implementing kernel density estimation. Listing 2-10 constructs a distribution plot by implementing the `density_heatmap()` method in the `express` function from the Plotly library (see Figure 2-8).

Listing 2-10. 2D Heatmap

```
figure = px.density_heatmap(df, x="cpi", y="rand",
                            title="South African consumer price index and
                            rand 2D histogram")
figure.show()
```

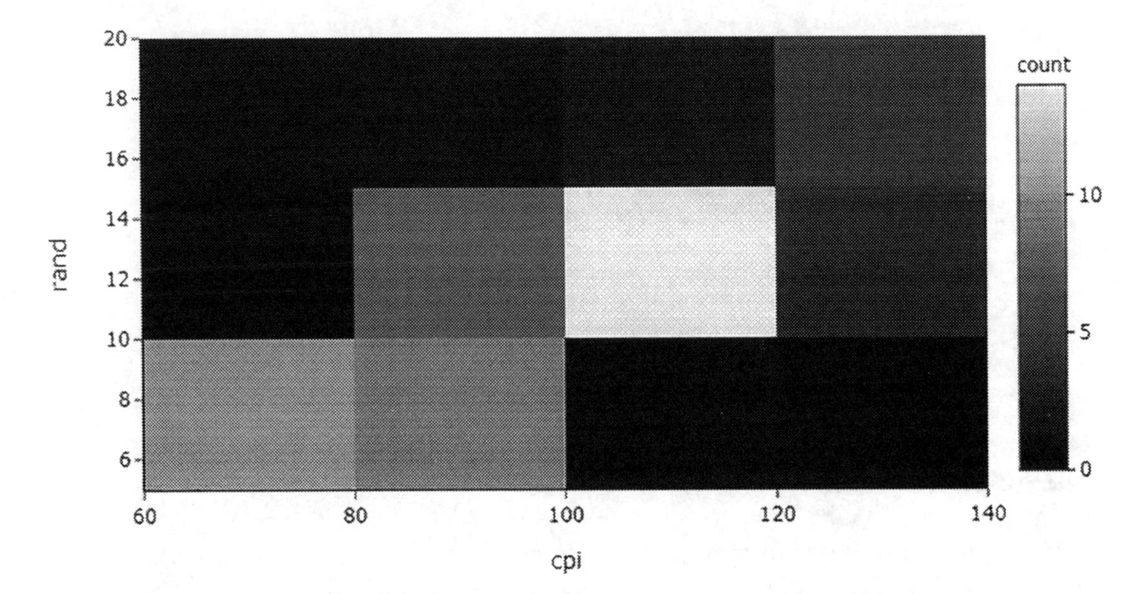

Figure 2-8. *2D heatmap*

Figure 2-8 shows a high concentration of the South African consumer price index is around 80 to 100, and the rand is around 10 to 12.

Distribution Plot

A distribution plot (also known as a *distplot*) combines many plots (e.g., histogram and kernel density estimation). Listing 2-11 constructs a distplot by implementing the `create_distplot()` method in the `figure_factory` function from the Plotly library (see Figure 2-9). First, it specifies the data and labels.

Listing 2-11. Distribution Plot

```
import plotly.figure_factory as ff
data = [df["m3"],df["rand"]]
labels = ["Money Supply", "Rand"]
figure = ff.create_distplot(data, labels)
figure.show()
```

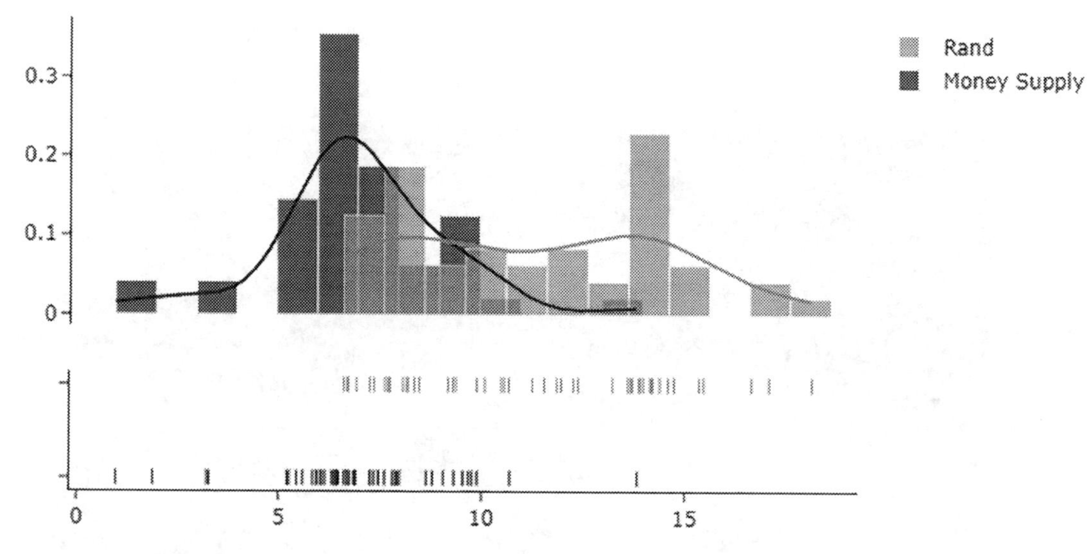

Figure 2-9. *Distribution plot*

Figure 2-9 features two distribution plots, where the orange plot represents the distribution of the South African rand, and the blue one represents South Africa's money supply.

Scatter Plot

Listing 2-12 constructs a scatter plot by implementing the `scatter()` method from the `express` function (see Figure 2-10).

Listing 2-12. Scatter Plot

```
figure = px.scatter(df, x="gdp_by_exp", y="rand",
                    title="South African GDP by expenditure and rand
                    scatter plot")
figure.show()
```

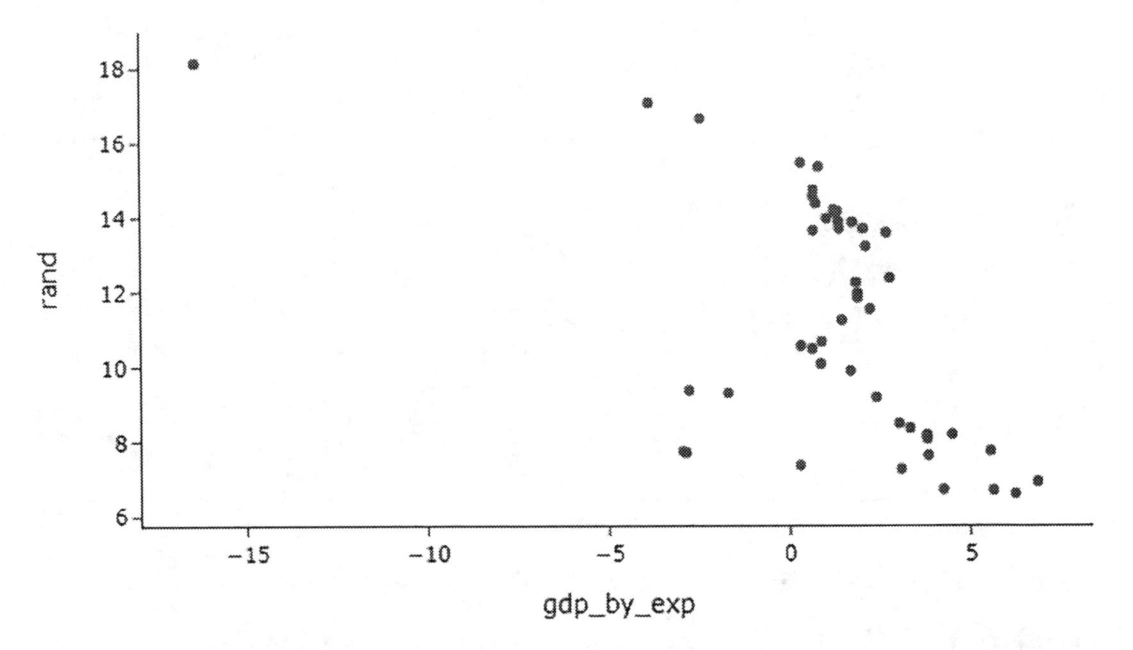

Figure 2-10. *Scatter plot*

Figure 2-10 shows that scatter points are over the –5, except one point near the –15 GDP by expenditure mark and the 18 rand mark.

Scatter Matrix

Instead of plotting the scatter in the data individually, you can plot all of them at once. Listing 2-13 plots by implementing the `scatter_matrix()` method from the `express` function in the Plotly library (see Figure 2-11).

Listing 2-13. Scatter Matrix

```
figure = px.scatter_matrix(df)
figure.show()
```

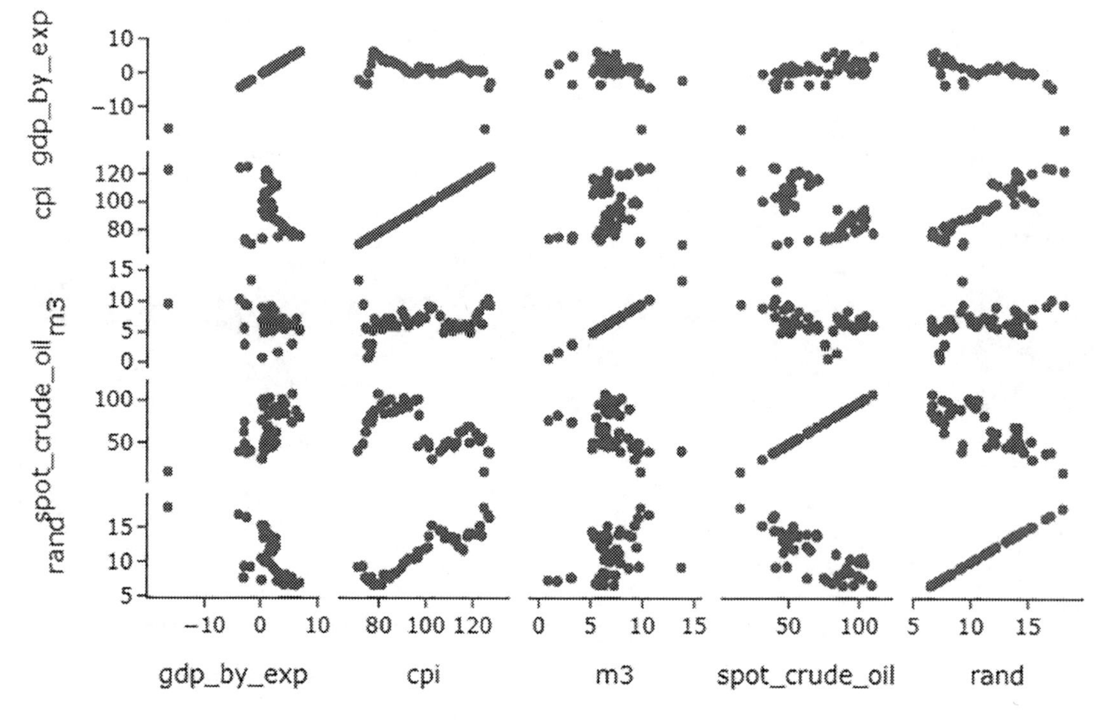

Figure 2-11. *Scatter matrix*

Figure 2-11 exhibits positive linear relationships (i.e., South African consumer price index and rand, and the rand and m3) and negative linear relationships (spot crude oil and rand).

Density Plot

Listing 2-14 constructs a density plot with "consumer price index" in the x-axis and "rand" in the y-axis by implementing the `scatter()` method from the `express` function (see Figure 2-12).

Listing 2-14. Density Plot

```
figure = px.density_contour(df, x="cpi", y="rand",
                            title="South African consumer price index and
                            rand density plot")
figure.show()
```

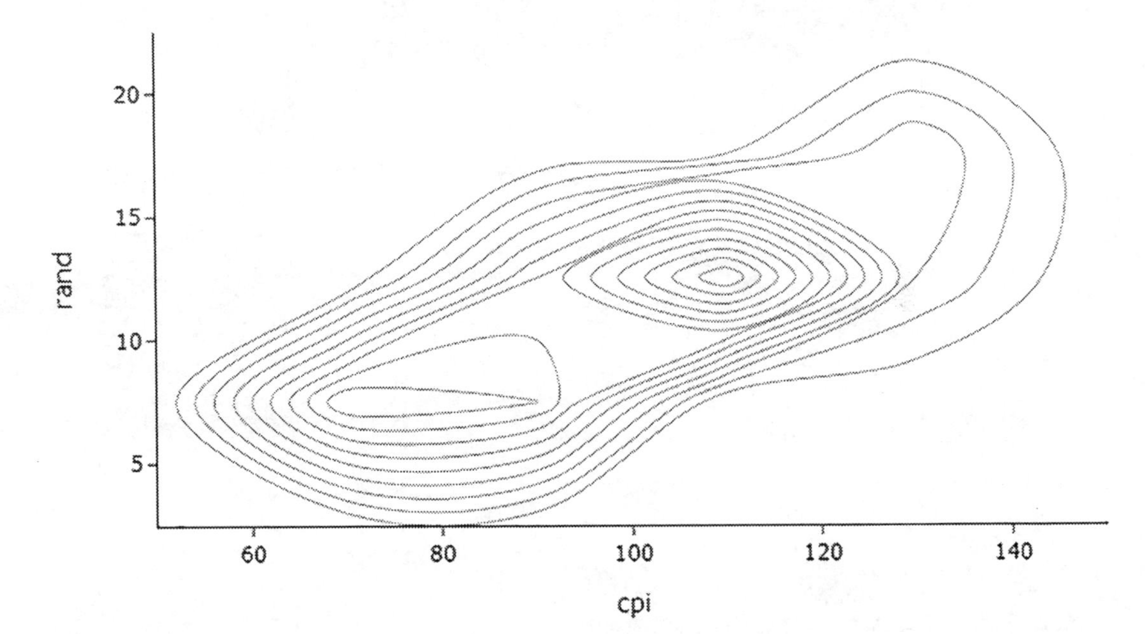

Figure 2-12. *Density plot*

In addition to the density plot, you can display the statistical distribution of each feature by specifying `marginal_x` and `marginal_y` (see Listing 2-15 and Figure 2-13).

Listing 2-15. Density Plot with Histogram

```
figure = px.density_contour(df, x="cpi", y="rand",
                            marginal_x="histogram", marginal_y="histogram")
figure.show()
```

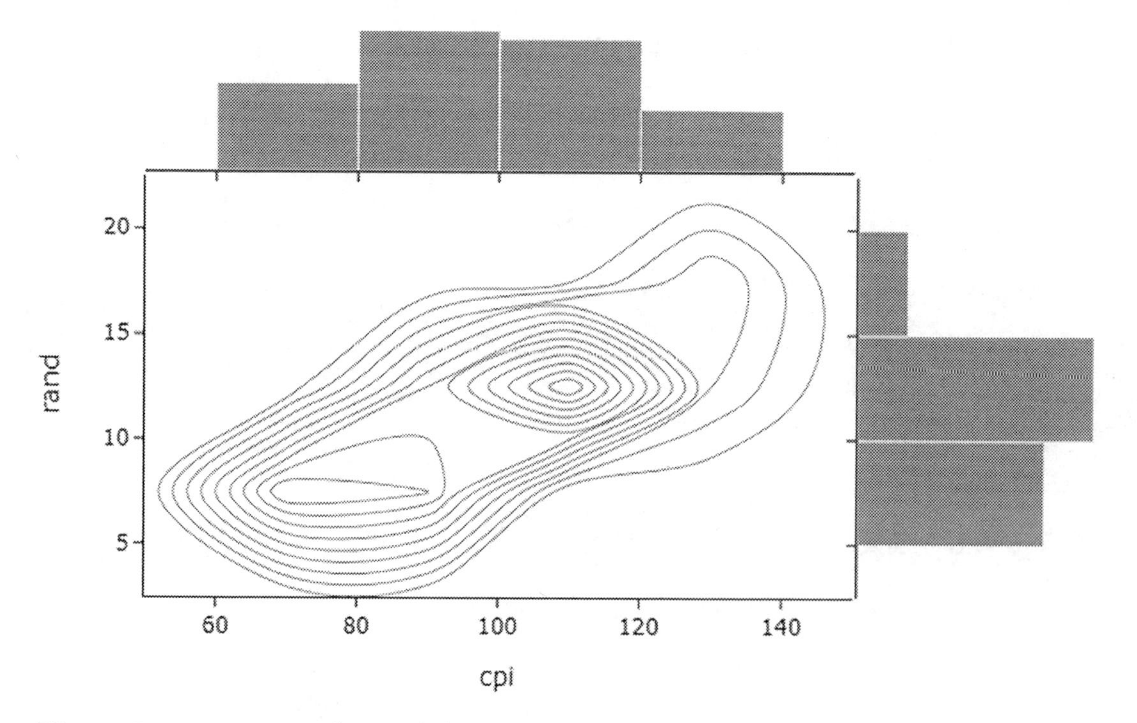

Figure 2-13. *Density plot with histogram*

Figure 2-13 shows that the South African consumer price index is close to being normally distributed. Meanwhile, the distribution of the rand is skewed to the left.

Bar Chart

A bar chart highlights the features on the x-axis and the count on the y-axis. The data was extracted from the World Bank database using the wbdata library. Install wbdata in a Python environment using `pip install wbdata`.

Listing 2-16 constructs a bar plot by implementing the `bar()` method from the `express` function (see Figure 2-14).

Listing 2-16. Bar Chart

```
import wbdata
g20_countries = ["ARG","AUS","BRA","CAN","CHN" ,
                 "FRA","DEU","IND" ,"IDN","ITA",
                 "JPN","MEX","RUS","SAU","ZAF",
                 "KOR","TUR","GBR","USA","EUU"]
```

```
economicInd = {"NY.GNP.PCAP.CD":"gni_per_capita"}
indicators = economicInd
gni_per_capita = wbdata.get_dataframe(indicators, country=g20_countries,
convert_date=True)
gni_per_capita_data = gni_per_capita["gni_per_capita"]
gni_per_capita_descr = pd.DataFrame(gni_per_capita)
gni_per_capita_descr = gni_per_capita_descr.reset_index()
gni_per_capita_dfu = gni_per_capita_data.unstack(level=0)
figure = px.bar(gni_per_capita_descr, x="country", y="gni_per_capita",
               title= "G20 countries GNI bar chart")
figure.show()
```

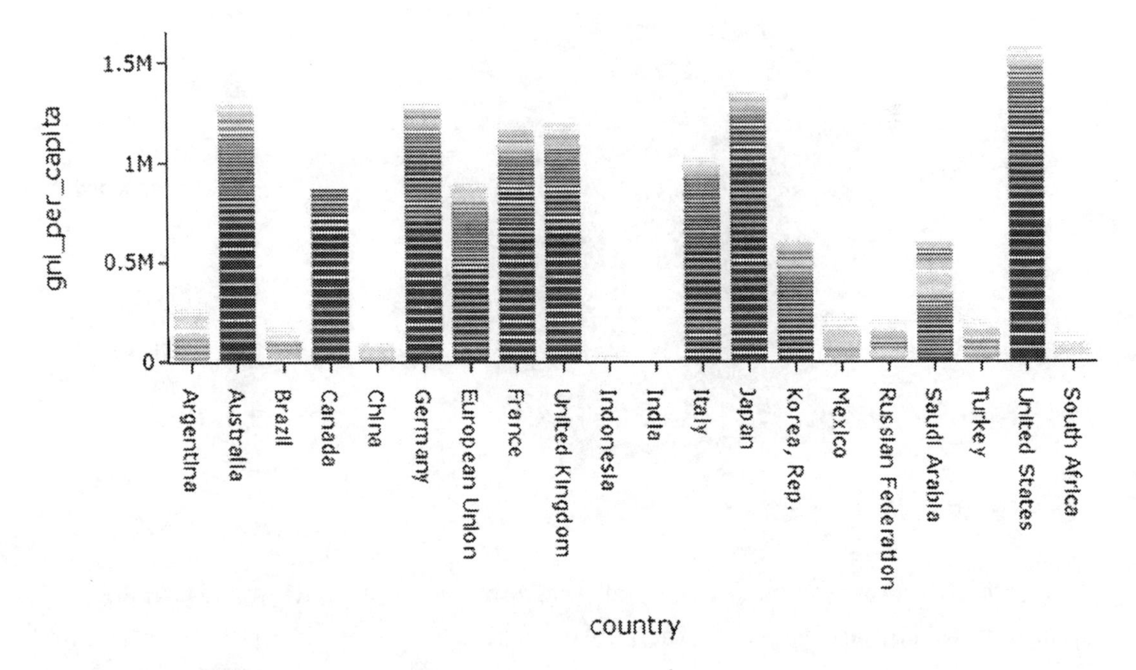

Figure 2-14. *Bar chart*

Figure 2-14 shows that the United States accounts the most for the world gross national income (GNI) per capita with more than $1.5 million, followed by Japan, Australia, Germany, France, the United Kingdom, and Italy with more than $1 million. The remaining countries fall below $1 million mark.

Pie Chart

A pie chart presents the percentage count per category. Listing 2-17 constructs a pie plot by implementing the `pie()` method from the `express` function (see Figure 2-15).

Listing 2-17. Pie Chart

```
figure = px.pie(gni_per_capita_descr, values="gni_per_capita",
names="country",
               title= "G20 countries GNI pie chart")
figure.show()
```

G20 countries GNI pie chart

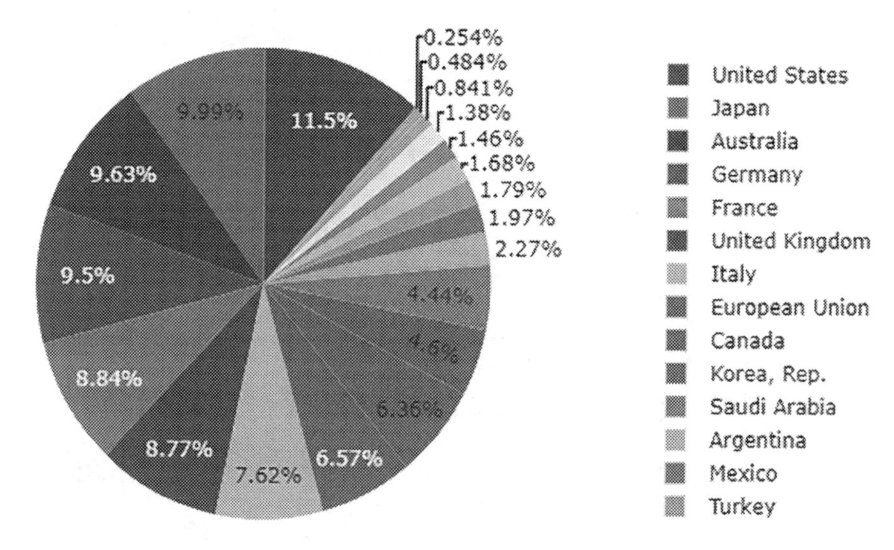

Figure 2-15. *Pie chart*

Figure 2-15 shows that the United States accounts for 11.5% of the world GNI per capita, followed by Japan at 9.99%, then Australia at 9.63%, and so forth.

Sunburst

A sunburst is convenient when there are several features because it simplifies them. Listing 2-18 constructs a sunburst by implementing the `sunburst()` method from the `express` function. Subsequently, it specifies the path as "`parent`" and "`labels`", hover_data as "values", and color_continuous_scale as "RdBu" (see Figure 2-16). The data was extracted from the World Bank database using the world-bank-data library. You can install it in a Python environment using pip install world-bank-data.

Listing 2-18. Sunburst

```
import world_bank_data as wb
countries_set = wb.get_countries()
gni_per_capita_int = pd.DataFrame(wb.get_series("NY.GNP.PCAP.CD", mrv=1))
gni_per_capita_int.columns = ["gni_per_capita"]
gni_per_capita_int = gni_per_capita_int.reset_index()
countries_set = wb.get_countries()
gni_per_capita = wb.get_series('AG.LND.AGRI.ZS', id_or_value='id',
simplify_index=True, mrv=1)
gni_per_capita_df = countries_set[['region', 'name']].
rename(columns={'name': 'country'}).loc[
    countries_set.region != 'Aggregates']
gni_per_capita_df['gni_per_capita'] = gni_per_capita
columns = ['parents', 'labels', 'values']
level1 = gni_per_capita_df.copy()
level1.columns = columns
level1['text'] = level1['values'].apply(lambda pop: '{:,.0f}'.format(pop))
level2 = gni_per_capita_df.groupby('region').gni_per_capita.sum().reset_
index()[['region', 'region', 'gni_per_capita']]
level2.columns = columns
level2['parents'] = 'World'
level2['text'] = level2['values'].apply(lambda pop: '{:,.0f}'.format(pop))
level2['values'] = 0
level3 = pd.DataFrame({'parents': [''], 'labels': ['World'],
                       'values': [0.0], 'text': ['{:,.0f}'.format(gni_per_
                       capita.loc['WLD'])]})
sunburst_gni_per_capita_all_levels = pd.concat([level1, level2, level3],
axis=0).reset_index(drop=True)
```

```
gni_per_capita_sunburst = px.sunburst(sunburst_gni_per_capita_all_levels,
path=['parents', 'labels'], values='values',
                                      color='text', hover_data=['values'],
                                      color_continuous_scale='RdBu')
gni_per_capita_sunburst.update_layout(margin=dict(t=10, l=10, r=10, b=10))
gni_per_capita_sunburst.show()
```

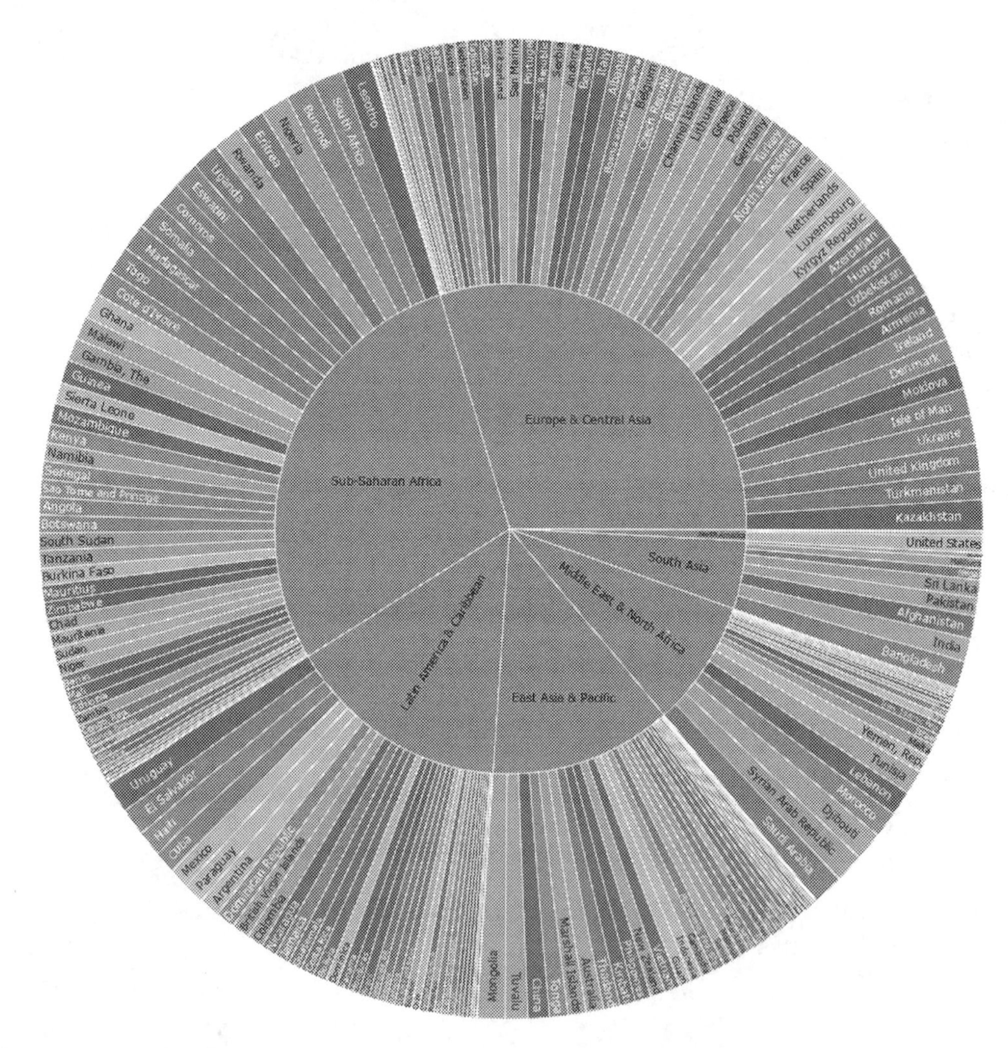

Figure 2-16. *Sunburst*

Choropleth Map

You want to use a choropleth map if you are dealing with an enormous amount of data from different countries or regions. A choropleth map presents data distribution across varying geographic boundaries. This section presents several countries' gross national income (GNI) distribution.

Listing 2-19 constructs a choropleth map (see Figure 2-17). First, it imports the `world_bank_data` library as `wb`, and then gets a list of all countries in the World Bank database. Subsequently, it implements the `get_series()` method from the `world_bank_data` library to get the series for each country. Following that, it specifies the country names and regions. Afterward, it implements a dictionary to specify the settings of the Plotly figure (`colorscale`, `reversescale`, `locations`, and `locationmode`, among others).

Listing 2-19. Choropleth Map

```
gni_per_capita_chro_data = dict(type = 'choropleth',
                               colorscale = 'Inferno',
                               reversescale = True,
                               locations = gni_per_capita_df['country'],
                               locationmode = "country names",
                               z = gni_per_capita_df['gni_per_capita'],
                               text = gni_per_capita_df['country'],
                               marker = dict(line = dict(color =
                               'rgb(255,255,255)',width = 1)),
                               colorbar = {'title' : ''})
gni_per_capita_map = go.Figure(data=[gni_per_capita_chro_data])
gni_per_capita_map.show()
```

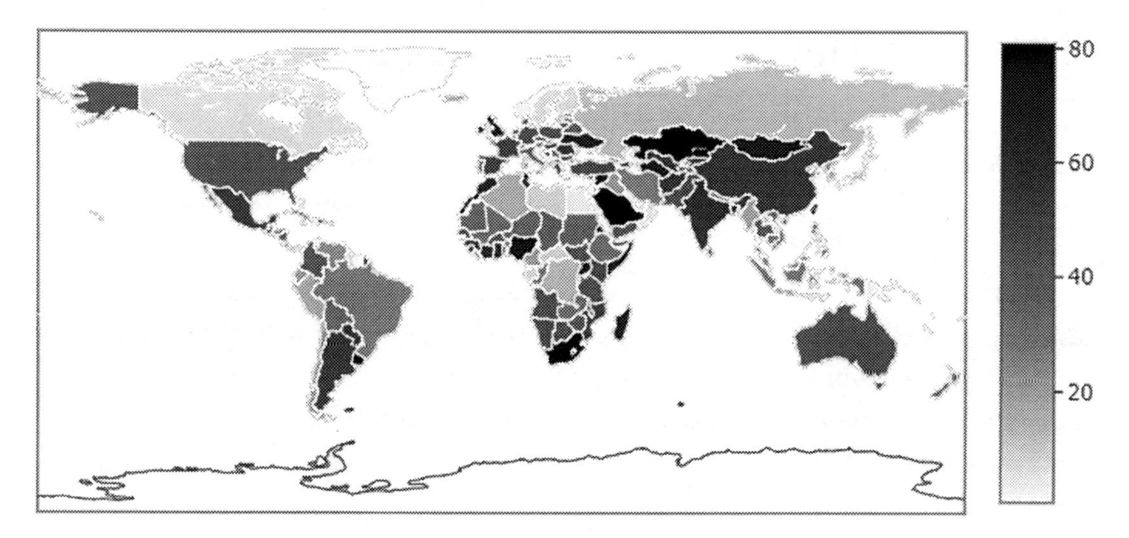

Figure 2-17. *Choropleth map*

Heatmap

A heatmap can be used to show the intensity of data within a certain range. Listing 2-20 contrasts a heatmap by implementing the `imshow()` method from the `express` function in Plotly (see Figure 2-18).

Listing 2-20. Heatmap

```
figure = figure = px.imshow(df,color_continuous_scale=px.colors.sequential.
Inferno,
                  title="South African economic data heatmap")
figure.show()
```

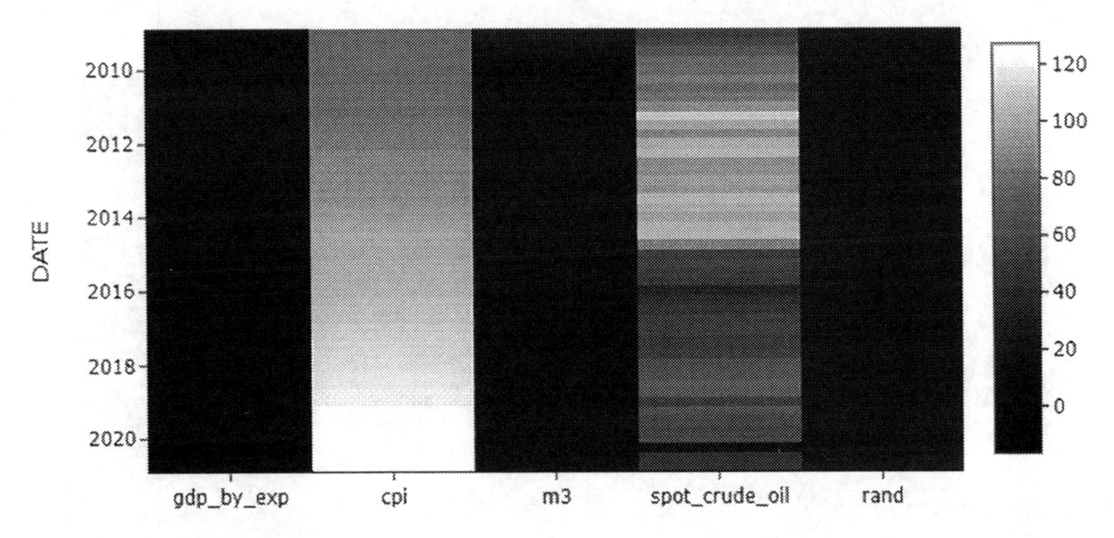

Figure 2-18. *Heatmap*

3D Charting

Alternatively, you may graphically represent data in a 3D space. Listing 2-21 constructs a 3D scatter plot that exhibits the relationship between "gdp_by_exp", "consumer price index", and "rand" by implementing the `scatter_3d()` method from the `express` function (see Figure 2-19).

Listing 2-21. 3D Scatter Plot

```
figure = px.scatter_3d(df, x="gdp_by_exp", y="cpi", z="rand",
                       color="rand",
                       title="South African GDP by expenditure, consumer
                       price index and rand 3D scatter plot")
figure.show()
```

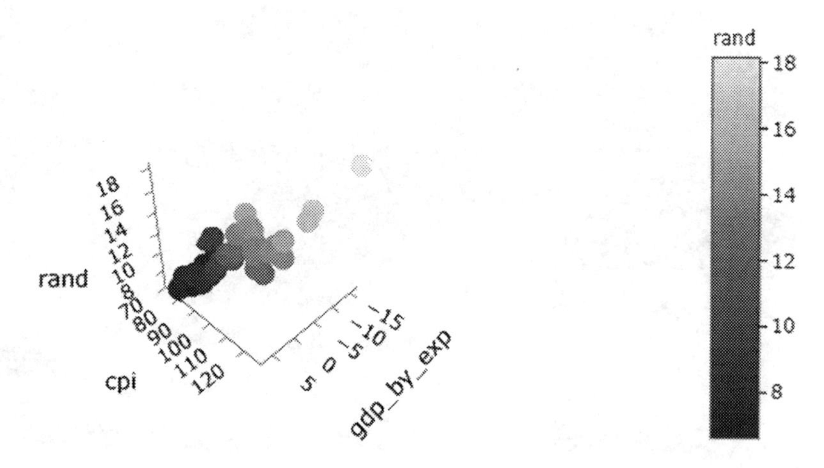

Figure 2-19. *Scatter plot*

Indicators

Except in the charts shown so far, indicators are used to simplify changes in the data. Indicators enable you to gauge the motion in the data. Listing 2-22 creates an indicator by implementing the `Indicator()` method from the `graph_objects` function in the Plotly library (see Figure 2-20). First, it finds the minimum value of the "rand" feature by implementing the `min()` method from pandas and the current value by implementing the `tail()` method and `iat[]`. Then, it specifies the value and reference of the delta.

Listing 2-22. Indicator

```
mininum_rand = df["rand"].min()
current_rand = pd.DataFrame(df["rand"].tail(2))
current_rand = df["rand"].tail(2).iat[-2]
fig_indicator = go.Figure()
fig_indicator.add_trace(go.Indicator(
    mode = "number+delta",
    value = current_rand,
    title = {"text": "South African rand<br><span style='font-
    size:0.8em;color:gray'>"},
```

```
    delta = {'reference': mininum_rand, 'relative': True},
    domain = {'x': [0, 1], 'y': [0.5, 1]}))
fig_indicator.show()
```

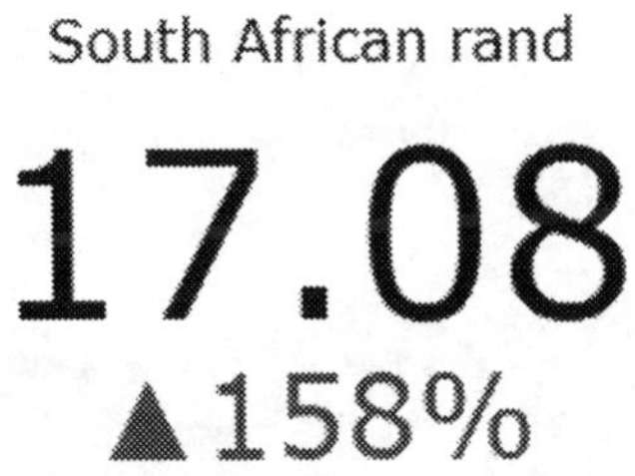

South African rand

17.08

▲158%

Figure 2-20. *Indicator*

Conclusion

This chapter acquainted you with constructing interactive 2D and 3D charts by implementing plotly, the most prevalent Python library. It presented basic charts (i.e., box-whisker plot, histogram, scatter plot, scatter matrix, density plot, heatmap, violin plot, sunburst, bar chart, pie chart, and choropleth map).

Chapter 3 introduces you to interactive functionalities created with the `express` or `graph_objects` functions.

CHAPTER 3

Containing Functionality and Styling for Interactive Charts

This chapter expands upon what was covered in Chapter 2. It introduces an approach to updating interactive graphs to improve user experience. For instance, you will learn how to add buttons and range sliders, among other functionalities. Additionally, this chapter describes how to integrate multiple graphs into one graph with functionality.

Updating Graph Layout

There are two main approaches to updating Plotly graphs. The first involves implementing the `update_layout()` method, and the second involves implementing `["layout"]` and specifying a string containing values that control the behavior of the graph. Listing 3-1 imports Plotly key dependencies and specify the template.

Listing 3-1. Import Plotly Key Dependencies

```
import plotly.express as px
from plotly.subplots import make_subplots
import plotly.io as pio
pio.templates.default = "simple_white"
```


T. C. Nokeri, *Web App Development and Real-Time Web Analytics with Python*,
https://doi.org/10.1007/978-1-4842-7783-6_3

Updating Plotly Axes

Plotly lets you update the x-axis and y-axis by implementing the update_xaxes() and update_yaxes() methods. Listing 3-2 updates both the x-axis and y-axis, including the title.

Listing 3-2. Update Graph Layout

```
figure = px.line(df, x=df.index, y="cpi",
                 title="South African consumer price index series")
figure.update_xaxes(title_text="Date ")
figure.update_yaxes(title_text="Consumer price index")
```

Implementing ["layout"] involves specifying a string with values for x, y, title, xaxis, yaxis, anchor, and domain. Listing 3-3 applies ["layout"] to update the x-axis and y-axis, title, and domain.

Listing 3-3. Update Graph Layout

```
figure = px.line(df,
                 x=df.index,
                 y="cpi")
figure["layout"] = {"title": "South African consumer price index series",
               "xaxis": {"anchor": "y", "domain": [0.0, 1.0],
                         "title": "Date"},
               "yaxis": {'anchor': "x", "domain": [0.0, 1.0],
                         "title": "Consumer price index"}}
```

Including Range Slider

A ranger slider enables users to select a range for view. Minimizing the range slider enables you to view observations of a short range. Listing 3-4 integrates a range slider into the graph by specifying rangeslider_visible as True (see Figure 3-1).

Listing 3-4. Including Range Slider

```
figure = px.line(df, x=df.index, y="rand", title="South African rand line plot")
figure.update_xaxes(rangeslider_visible=True)
figure.show()
```

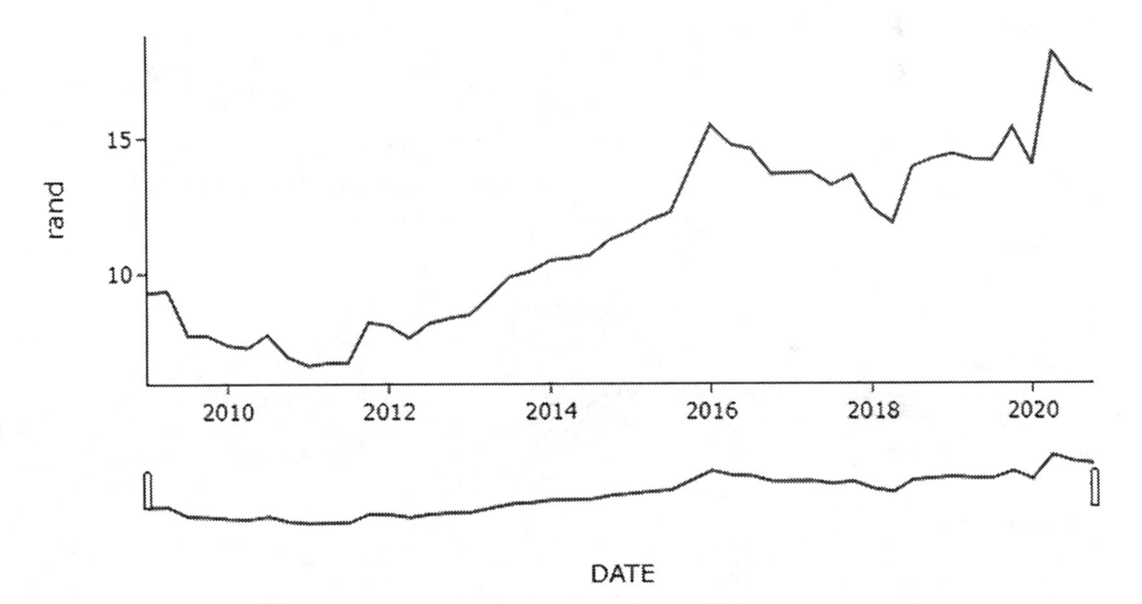

Figure 3-1. *Line plot with a range slider*

Figure 3-1 exhibits a series with a range slider that enables the user to zoom in and out of a chart, thus bounding the series to a certain date range.

Including Buttons to a Graph

You can add a button to the graph for some functionality. Listing 3-5 adds buttons to select a period range by implementing the `update_layout()` method. Where `"1m"` represents one month, `"6m"` represents six months, `"1yr"` represents one year, `"YTD"` represents today's date. To specify the period, use `step` (i.e., `"day"`, `"week"`, `"month"`, or `"year"`) and specify `stepmode` as "`backward`" (see Figure 3-2).

Listing 3-5. Including Buttons to a Graph

```
figure = px.line(df, x=df.index, y="cpi",
                title="South African consumer price index series")
figure.update_xaxes(title_text="Date ")
figure.update_yaxes(title_text="Consumer price index")
figure.update_xaxes(rangeslider_visible=True,
                    rangeselector=dict(
                        buttons=Listing([
                            dict(count=1, label="1m",
                                 step="month", stepmode="backward"),
                            dict(count=6, label="6m",
                                 step="month", stepmode="backward"),
                            dict(count=1, label="YTD",
                                 step="year",
                                 stepmode="todate"),
                            dict(count=1, label="1y",
                                 step="year",
                                 stepmode="backward"),
                            dict(step="all")])))
figure.show()
```

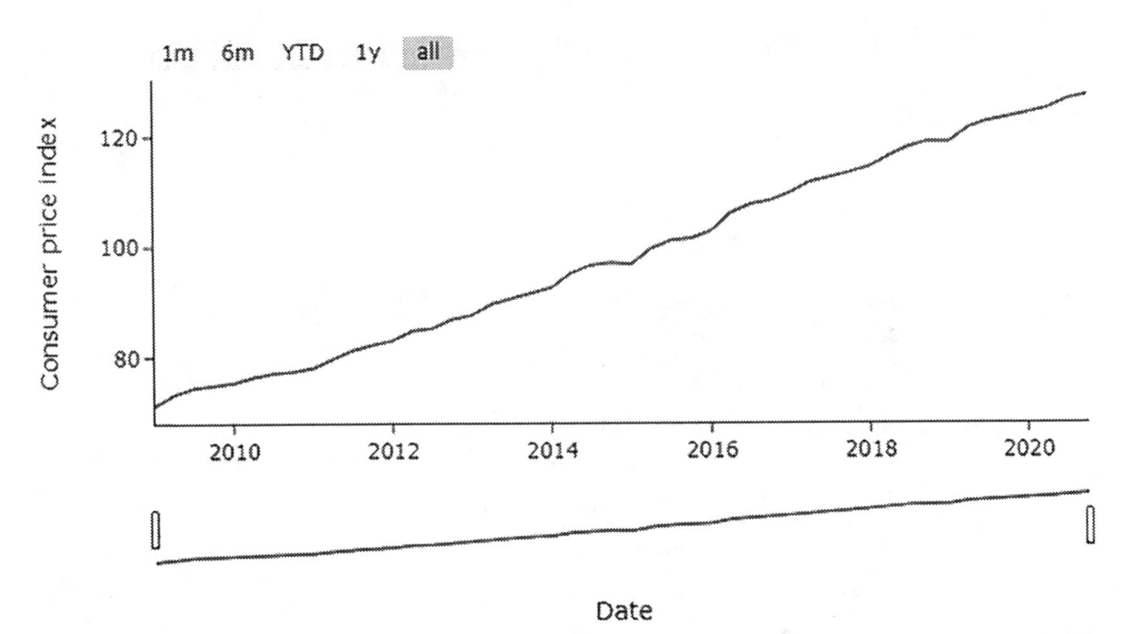

Figure 3-2. *Line plot with a range slider*

Figure 3-2 features a series with several buttons. The 1m button limits the series to one month, the 6m button limits it to six months, the 1yr button limits it to one year, and the YTD `button` limits it to today's date.

Subplots

To make subplots, implement the `make_subplots()` method from the `subplots` function in the Plotly library. The example beneath integrates three line plots. In addition to that, it comprises buttons that take the user to a specific line plot.

At times, you may want to incorporate different plots into one figure. When implementing the Plotly library, use the `make_subplot()` method from the `subplots` function. To achieve this, you must specify `rows` and `cols` (columns) each chart belongs to.

Listing 3-6 constructs a chart with multiple traces (see Figure 3-3).

Listing 3-6. Tracing in a Single Chart

```
from plotly.subplots import make_subplots
import plotly.graph_objects as go
figure = make_subplots(rows=1,
                       cols=1)
figure.add_trace(go.Scatter(x=df.index,
                            y=df.cpi,
                            name="Consumer price index",
                            mode="lines"),
                 row=1,col=1)
figure.add_trace(go.Scatter(x=df.index,
                            y=df.gdp_by_exp,
                            name="GDP per expenditure",
                            mode="lines"),
                 row=1,col=1)
figure.add_trace(go.Scatter(x=df.index,
                            y= df.rand,
                            mode="lines",
                            name="Rand",),
                 row=1,col=1)
figure.update_layout(paper_bgcolor = "rgba(0,0,0,0)",
                     plot_bgcolor = "rgba(0,0,0,0)",
                     showlegend=True)
figure.update_yaxes(title_text="Consumer price index", row=1,col=1)
figure.update_xaxes(title_text="Date", row=1,col=1)
figure.update_yaxes(title_text="Rand", row=1,col=1)
figure.show()
```

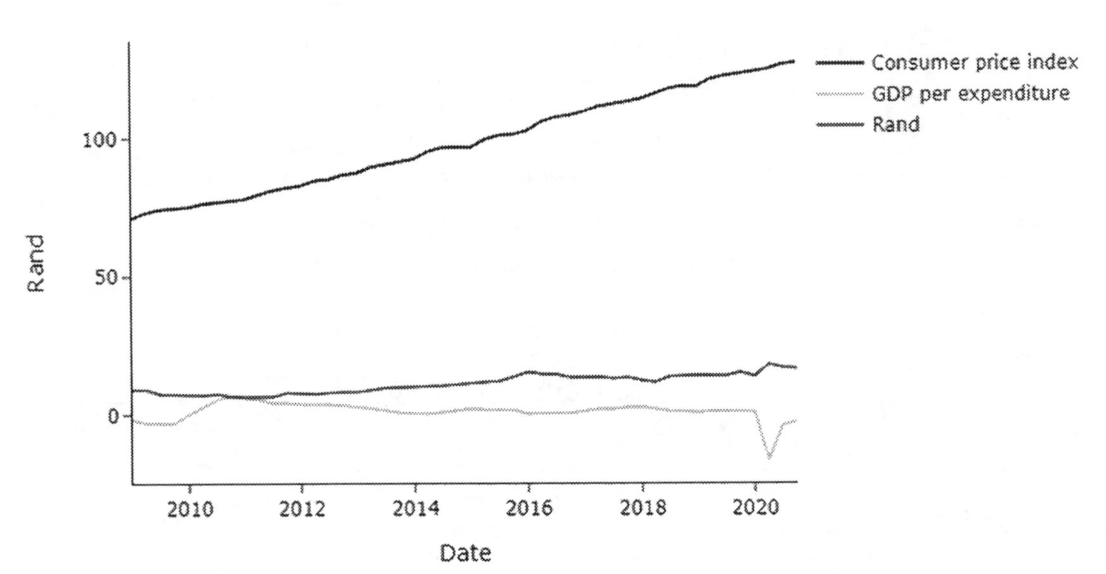

Figure 3-3. *Tracing in a single chart*

Figure 3-3 shows a series with several few traces (i.e., consumer price index, GDP by expenditure, and rand) with varying colors.

Listing 3-7 constructs multiple charts on top of each other (see Figure 3-4).

Listing 3-7. Creating Subplots

```
figure = make_subplots(rows=2,
                       cols=1,
                       row_heights=[0.7, 0.3],
                       subplot_titles=("Consumer price index",
                                       "Rand"))
figure.add_trace(go.Scatter(x=df.index,
                            y=df.cpi,
                            name="Consumer price index",
                            mode="lines"),
                 row=1,col=1)
figure.add_trace(go.Scatter(x=df.index,
                            y= df.rand,
                            mode="lines"),
                 row=2,col=1)
```

```
figure.update_layout(paper_bgcolor = "rgba(0,0,0,0)",
                     plot_bgcolor = "rgba(0,0,0,0)",
                     showlegend=False)
figure.update_yaxes(title_text="Consumer price index", row=1,col=1)
figure.update_xaxes(title_text="Date", row=2,col=1)
figure.update_yaxes(title_text="Rand", row=2,col=1)
figure.show()
```

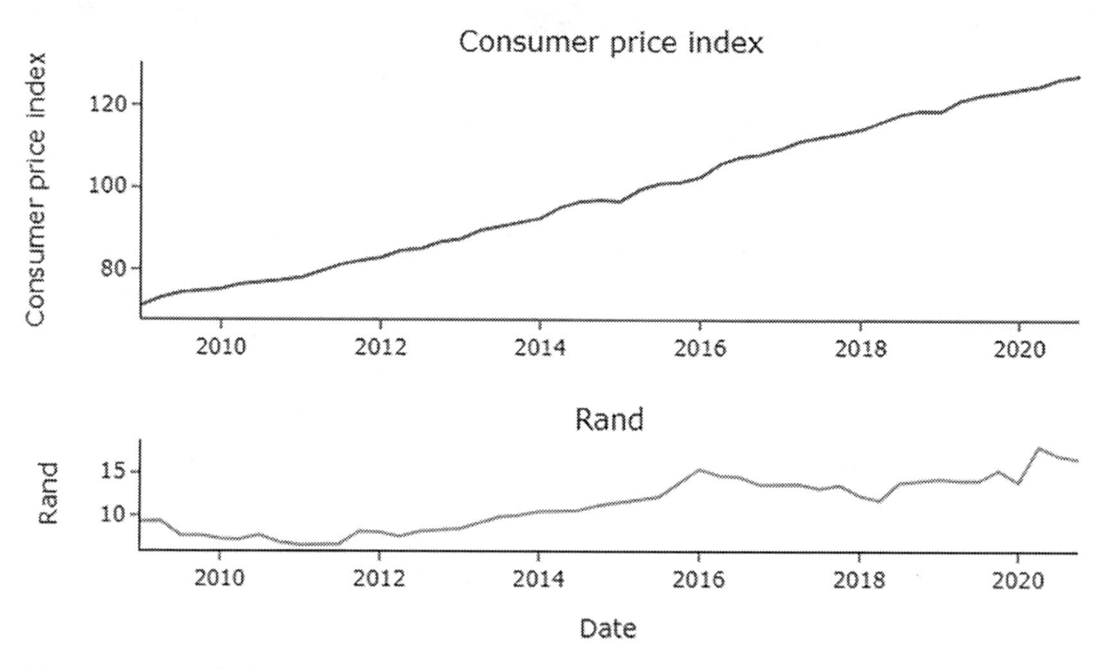

Figure 3-4. *Subplots*

Figure 3-4 presents two subplots. The top subplot relates to the consumer price index series, and the one beneath it relates to the rand series.

Listing 3-8 constructs a chart with multiple side-by-side (see Figure 3-5).

Listing 3-8. Creating Subplots

```
figure = make_subplots(rows=1,
                       cols=2,
                       subplot_titles=("Consumer price index",
                                       "Rand"))
```

```
figure.add_trace(go.Scatter(x=df.index,
                            y=df.cpi,
                            name="Consumer price index",
                            mode="lines"),
                 row=1,col=1)
figure.add_trace(go.Scatter(x=df.index,
                            y= df.rand,
                            mode="lines"),
                 row=1,col=2)
figure.update_layout(paper_bgcolor = "rgba(0,0,0,0)",
                     plot_bgcolor = "rgba(0,0,0,0)",
                     showlegend=False)
figure.update_yaxes(title_text="Consumer price index", row=1,col=1)
figure.update_xaxes(title_text="Date", row=1,col=1)
figure.update_yaxes(title_text="Rand", row=1,col=2)
figure.update_xaxes(title_text="Date", row=1,col=2)
figure.show()
```

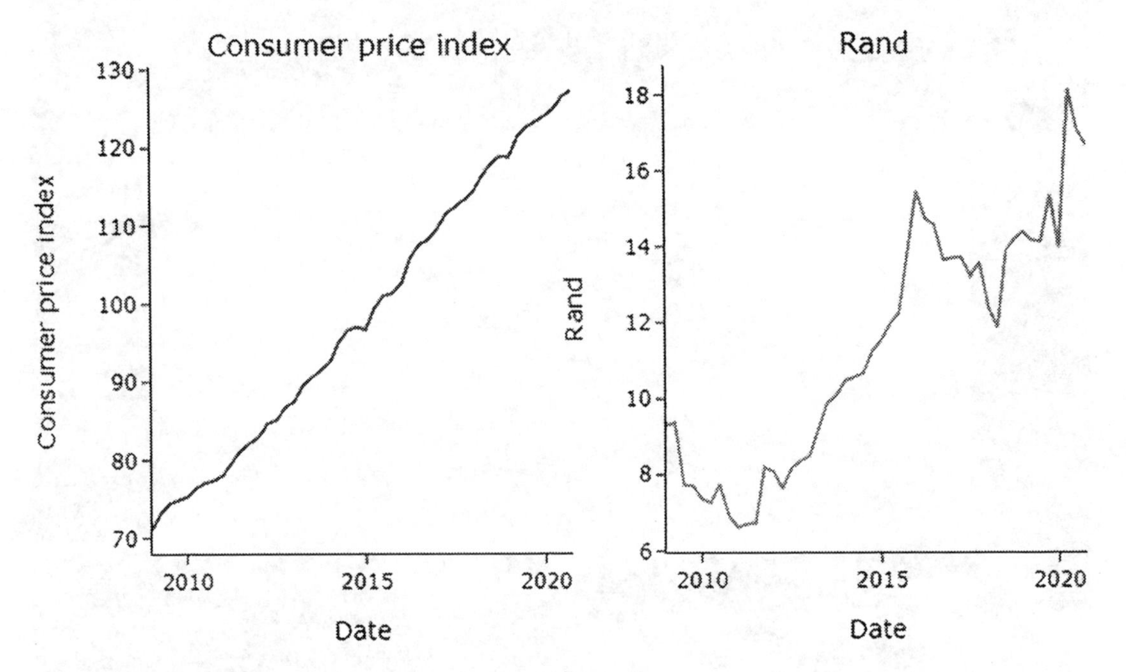

Figure 3-5. *Subplots*

Figure 3-5 exhibits a figure with two subplots side-by-side; the subplot on the left relates to the consumer price index series and the one on the right relates to the rand series.

Styling Charts

To alter the color, specify `plot_bgcolor` and `paper_bgcolor`. Listing 3-9 alters the plot background color and paper background color by implementing the `update_layout()` method from the Plotly library (see Figure 3-6).

Listing 3-9. Background Color and Paper Background Color

```
figure = px.line(df, x=df.index, y="cpi",
                 title="South African consumer price index series")
figure.update_layout(plot_bgcolor= "white",
                     paper_bgcolor= "gray")
figure.show()
```

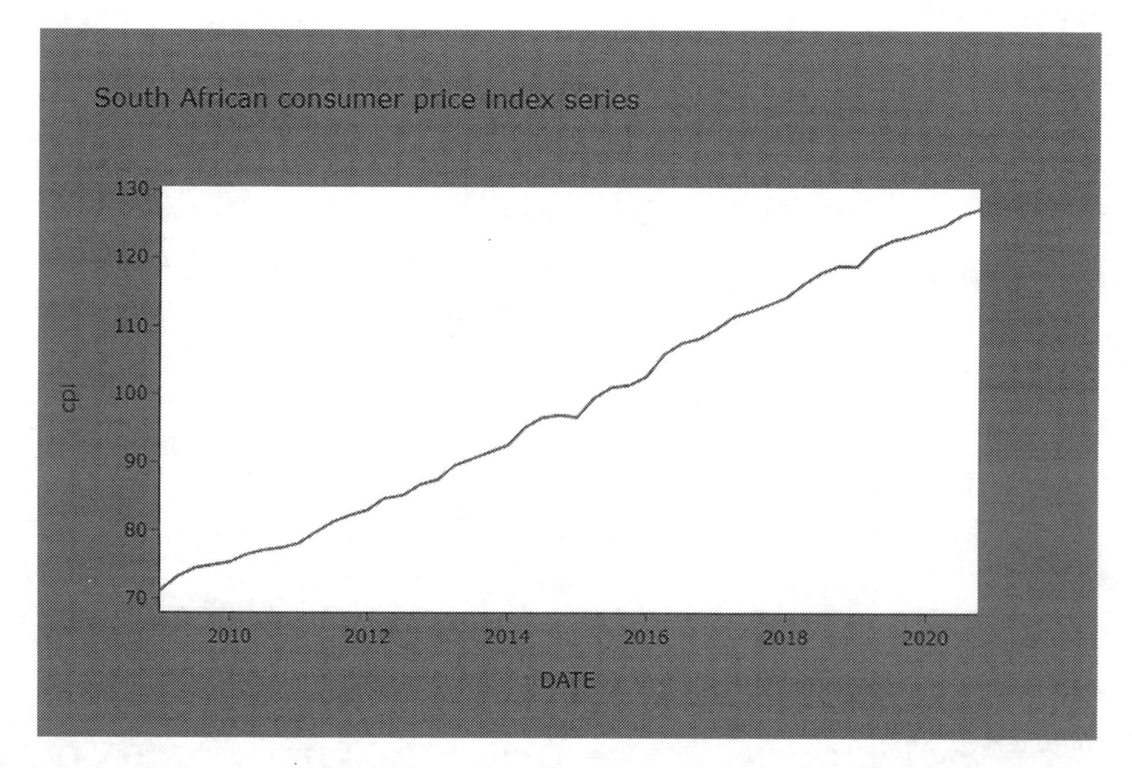

Figure 3-6. *Graph with black plot background and orange paper background*

Altering Color Schemes

You can use hex color code or RGB color coding to alter the colors of traces in a chart. For instance, using the hex color code, you can specify a color as "#757575", or using the RGB color coding style, you can specify it as "rgb(117, 117, 117)".

Color Sequencing

You can change the color sequence of data points in a graph. By doing so, the color changes as data points change. Listing 3-10 incorporates a color sequence called Inferno in a histogram by specifying color_continuous_scale (see Figure 3-7).

Listing 3-10. Incorporating a Color Sequence

```
figure = px.imshow(df,color_continuous_scale=px.colors.sequential.Inferno,
                   title="South African economic data heatmap")
figure.show()
```

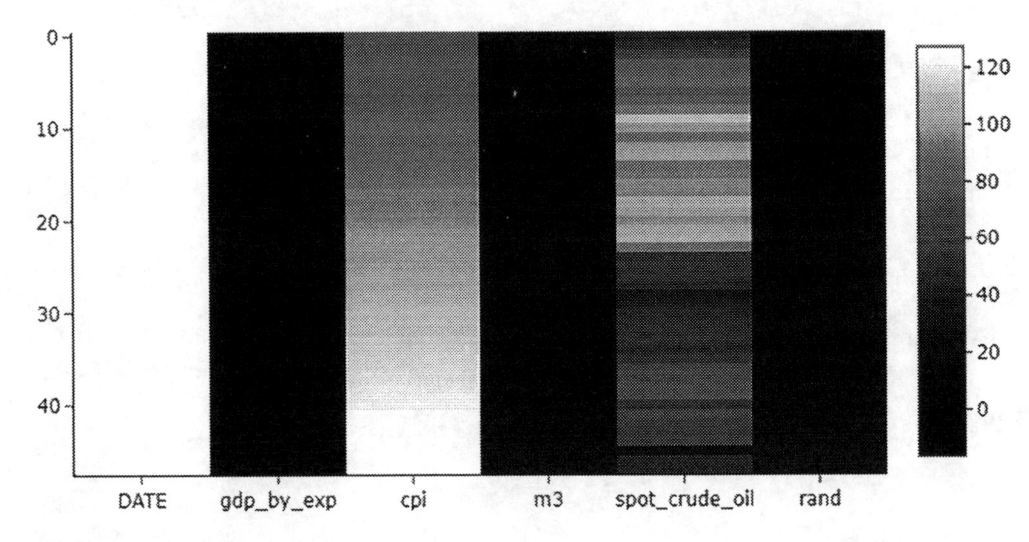

Figure 3-7. *Scatter plot with a color sequence*

Listing 3-11 incorporates a color sequence called Viridis into a histogram by specifying it in color_continuous_scale (see Figure 3-8).

Listing 3-11. Incorporating a Color Sequence

```
figure = px.imshow(df,color_continuous_scale=px.colors.sequential.Viridis,
                   title="South African economic data heatmap")
figure.show()
```

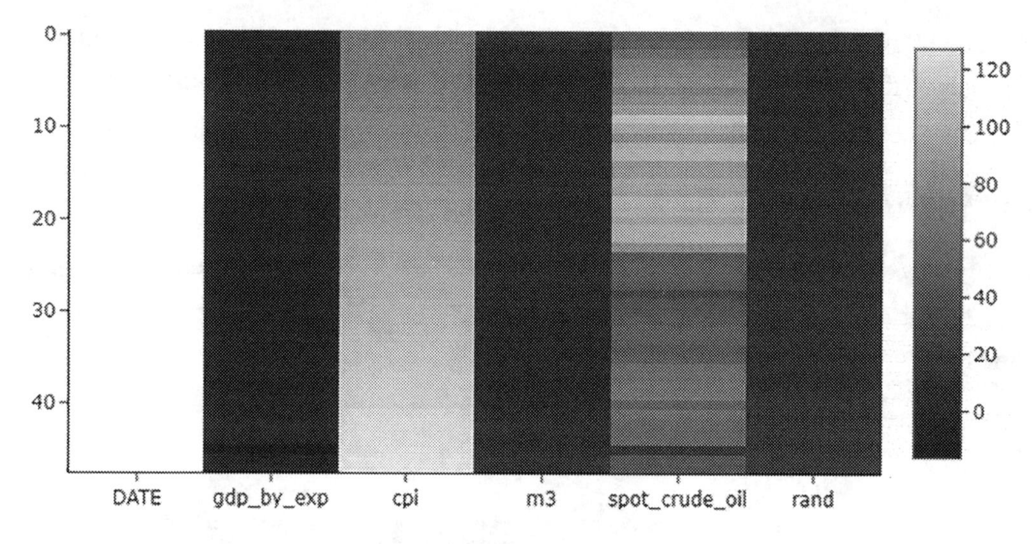

Figure 3-8. *Scatter plot with a color sequence*

Listing 3-12 incorporates a color sequence called `Plotly3` into a histogram by specifying it in `color_continuous_scale` (see Figure 3-9).

Listing 3-12. Incorporating a Color Sequence

```
figure = px.imshow(df,color_continuous_scale=px.colors.sequential.Plotly3,
                   title="South African economic data heatmap")
figure.show()
```

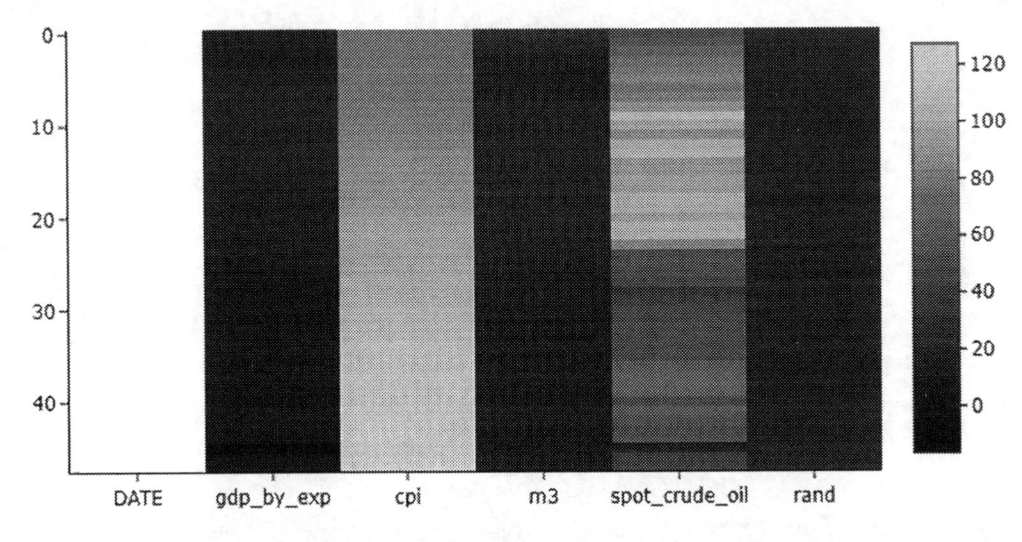

Figure 3-9. *Scatter plot with a color sequence*

There are color sequences besides the ones mentioned (i.e., dark mint, electric, and gray, among others). Learn more on the official Plotly website at `https://plotly.com/python/builtin-colorscales/`.

Customizing Traces

You may want to customize traces in a chart to improve the appeal. The easiest way involves manipulating `fillcolor` (see Figure 3-10).

Listing 3-13. Customizing Traces

```
figure = make_subplots(rows=2,
                       cols=1,
                       row_heights=[0.7, 0.3],
                       subplot_titles=("Consumer price index",
                                       "Rand"))
figure.add_trace(go.Scatter(x=df.index,
                            y=df.cpi,
                            name="Consumer price index",
                            mode="lines",
```

```
                            fill="tonexty",
                            fillcolor = "#1266F1",
                            line = dict(color = "#1266F1")),
                row=1,col=1)
figure.add_trace(go.Scatter(x=df.index,
                            y= df.rand,
                            mode="lines",
                            fill="tonexty",
                            fillcolor =  "rgb(117, 117, 117)",
                            line = dict(color = "rgb(255, 111, 0)")),
                row=2,col=1)
figure.update_layout(paper_bgcolor = "rgba(0,0,0,0)",
                     plot_bgcolor = "rgba(0,0,0,0)",
                     showlegend=False)
figure.update_yaxes(title_text="Consumer price index", row=1,col=1)
figure.update_xaxes(title_text="Date", row=2,col=1)
figure.update_yaxes(title_text="Rand", row=2,col=1)
figure.show()
```

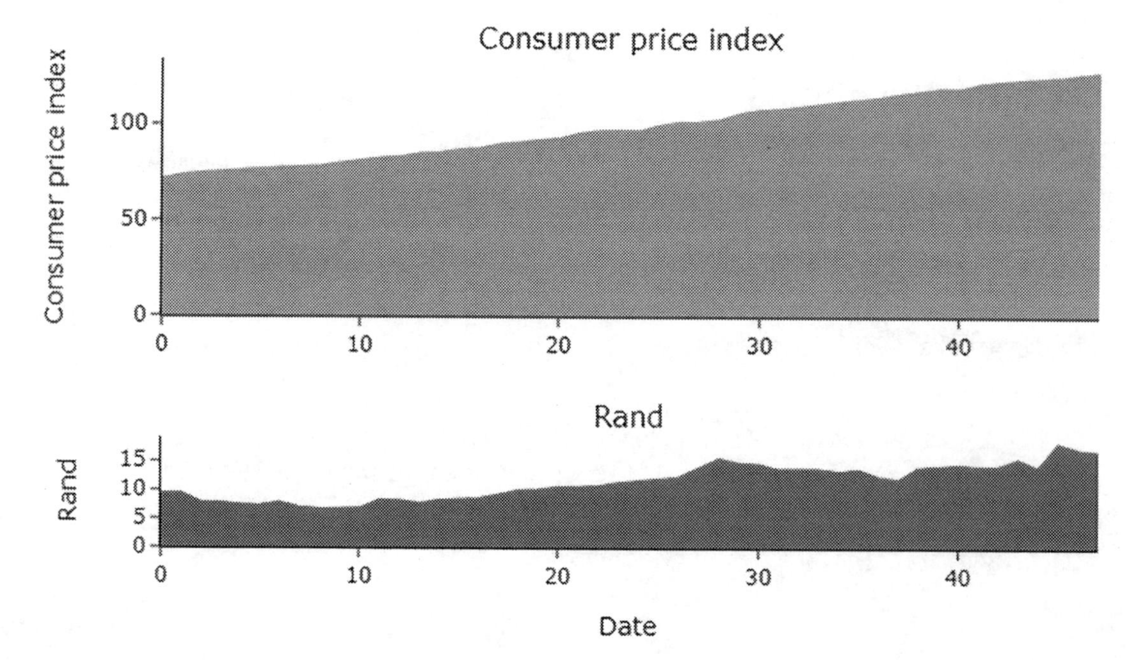

Figure 3-10. *Customizing traces*

Conclusion

This chapter presented a way to update interactive charts. It introduced updating a graph's layout, axis, and color sequence and demonstrated how to include buttons on a chart to enable response.

CHAPTER 4

Essentials of HTML

This chapter introduces the most widely used markup language for developing websites, and acquaints you with the essentials of designing websites. The examples included in this chapter will support you in getting started with HTML.

Communication Between a Web Browser and a Web Server

Web users with an Internet connection can access the contents of websites through a web browser, which initiates a HyperText Transfer Protocol (HTTP) session by establishing Transmission Control Protocol (TCP) to a specific port on a web server. Subsequently, the web server responds to the request with the content of a web page that the browser renders. Figure 4-1 illustrates the basic communication between a web browser and a web server.

Figure 4-1. *Basic structure of websites*

URL Structure

Each website has its own domain name (which is a unique name that is attached to a certain Internet Protocol (IP) address) that distinguishes it from other web pages (i.e., apress). And, each website has its own domain name extension (i.e., .com) (see Figure 4-2).

T. C. Nokeri, *Web App Development and Real-Time Web Analytics with Python*,
https://doi.org/10.1007/978-1-4842-7783-6_4

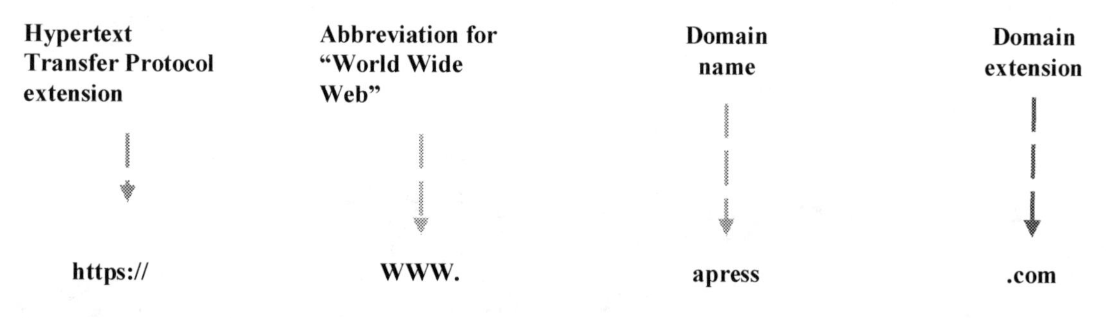

Figure 4-2. *URL structure*

The S in HTTPS stands for *secure*. It indicates the domain consists of a Secure Sockets Layer/Transport Layer Security (SSL/TLS) Wildcard certificate that secures the domain and its subdomains. The prevalent issuer of SSL/TLS Wildcard certificates is DigiCert (`www.websecurity.digicert.com/security-topics/what-is-ssl-tls-https`).

Domain Hosting

The most prevalent way of deploying websites involves outsourcing services from third-party companies. These companies managed web services that enable developers to deploy websites over the Internet using a URL. Platforms offer two services: shared web hosting and managed web hosting. The next section distinguishes shared and managed web hosting for background and perspective.

Shared Web Hosting

Shared web hosting involves sharing web server resources and services with other domains. For instance, assuming six domains are sharing a web server with a 6 gigabyte (GB) storage space and 6 RAM (random access memory), each domain has 1 GB storage space and 1 GB RAM allotted to it. Figure 4-3 simplifies shared web hosting.

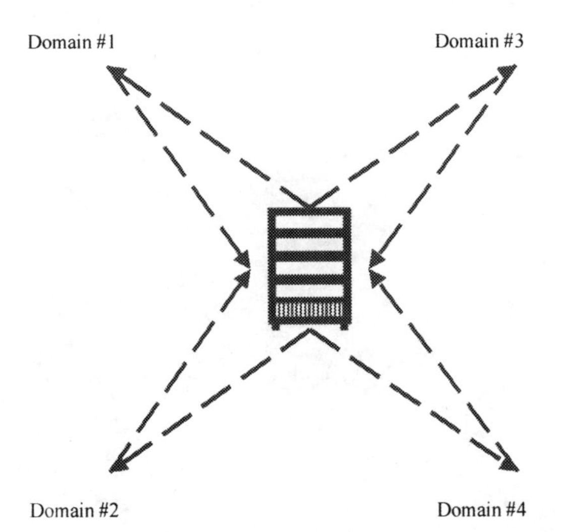

Figure 4-3. *Shared web hosting*

Figure 4-3 shows that in shared web hosting, a domain shares web server resources with other domains.

Managed Web Hosting

Alternatively, you may opt for managed web hosting, whereby a domain has full access to web server resources and services. Using the prior example, this entails that a domain has the full 6 GB storage space and 6 GB RAM. Developers can also manage resources and services on their own. Figure 4-4 simplifies shared web hosting.

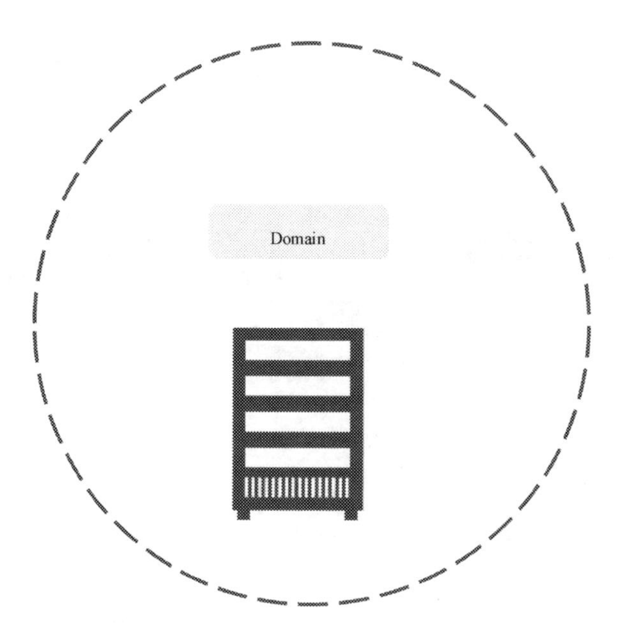

Figure 4-4. *Managed web hosting*

Figure 4-4 illustrates managed web hosting reserves web server resources and services for a specific domain.

Web Servers

A web server is an application that enables developers to deploy websites through a URL. It responds to requests that web browsers make over the Internet by channeling the connection through ports. Two of the most prevalent web servers are Windows Server and Linux servers. These servers store and manage HTML documents and supporting files.

Now that you have some background and perspective, let's explore the core of this chapter, which is the essentials of HTML.

HyperText Markup Language

HyperText Markup Language (HTML) is the most prevalent language for composing web pages. You can apply with other languages and frameworks to attain optimal results.

HTML Elements

Listing 4-1 exhibits the basic structure of an HTML file. `<!DOCTYPE html>` specifies that it is an HTML file. All HTML code starts with `<html>` and concludes with `</html>`.

Listing 4-1. Basic HTML

```
<!DOCTYPE html>
<html>
<title> Apress About Us <title>
<body> </body>
</html>
```

All elements begin with an opening tag (<>) and conclude with a closing tag (</>). Figure 4-5 illustrates these HTML elements and the usage of tags.

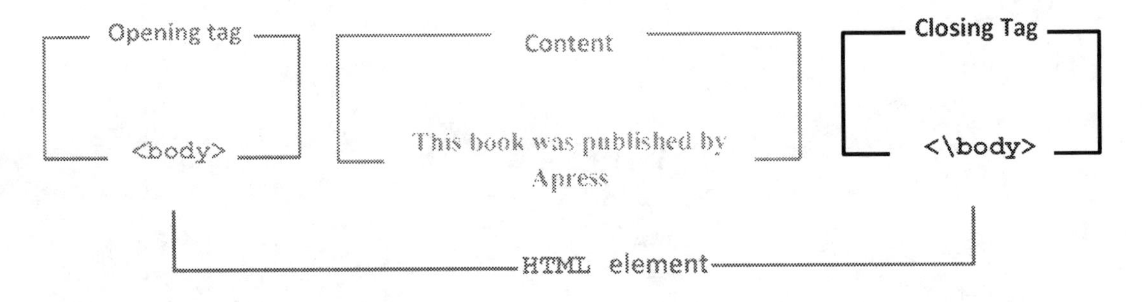

Figure 4-5. *HTML elements*

Figure 4-5 shows that the HTML element begins with the opening tag, `<body>`, then specifies the content of the body and concludes by stipulating the closing tag `</body>`.

Headings

Headings come in different sizes. Listing 4-2 constructs headings (see Figure 4-6).

Listing 4-2. HTML Headings

```
<!DOCTYPE html>
<html>
<h1>Apress</h1>
<h2>Apress</h2>
<h3>Apress</h3>
<h4>Apress</h4>
<h5>Apress</h5>
<h6>Apress</h6>
</html>
```

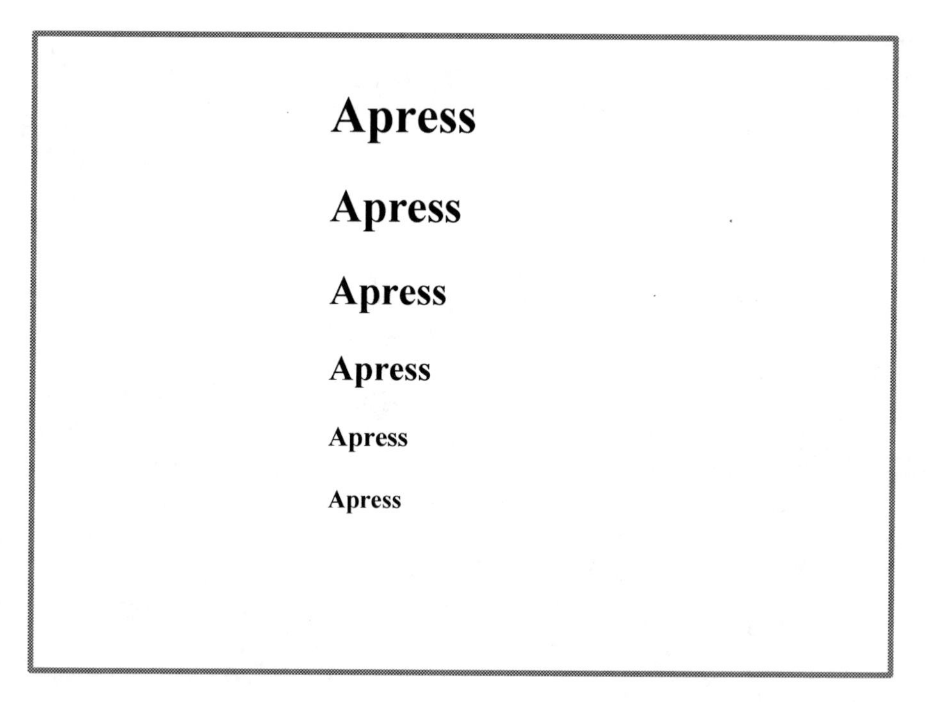

Figure 4-6. *HTML headings*

Paragraphs

You can write paragraphs in the body. Listing 4-3 constructs headings (see Figure 4-7).

Listing 4-3. HTML Headings and Paragraphs

```
<!DOCTYPE html>
<html>
<body>
<h1>Submit a book to Apress</h1>
<p>Becoming an Apress Author. Apress is looking for authors with both
technical expertise and the ability to clearly explain complicated
technical concepts. We want authors who are passionate, innovative, and
original. </p>
</body>
</html>
```

Submit a book to Apress

Becoming an Apress Author. Apress is looking for authors with both technical expertise and the ability to clearly explain complicated technical concepts. We want authors who are passionate, innovative, and original.

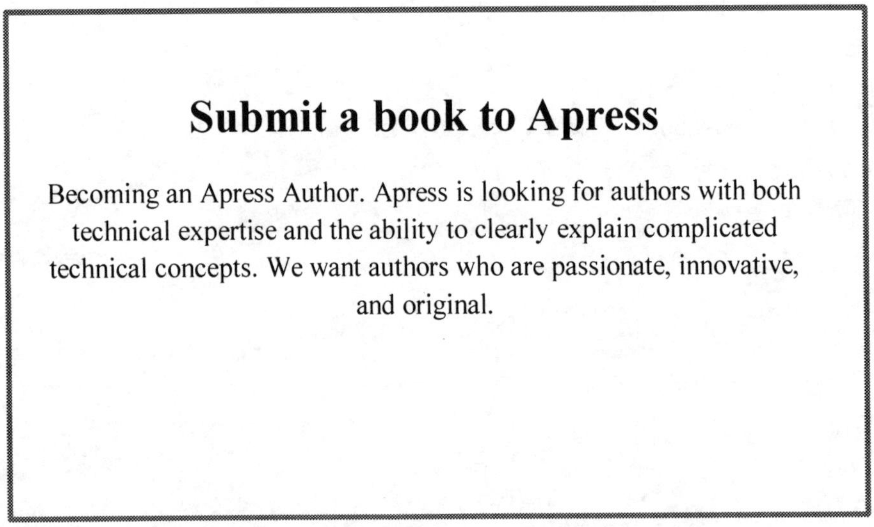

Figure 4-7. *HTML headings and paragraphs*

Div

`div` enables one to create a section for styling. Listing 4-4 constructs `div`. Notice that it applies `<style> </ style>` to set a div with `5px outset gray`, a `background-color` that is white and `text-align` is `center` (see Figure 4-8).

Listing 4-4. Div

```
<!DOCTYPE html>
<html>
<head>
<style>
.div_1{
  border: 5px outset gray;
  background-color: white;
  text-align: center;
}
</style>
</head>
<body>

<div class="div_1">
  <h2>Real-Time Dashboards and Web Apps with Python</h2>
  <p>A book brought to you by Apress</p>
</div>
</body>
</html>
```

Real-Time Dashboards and Web Apps with Python

A book brought to you by Apress

Figure 4-8. *HTML div*

Span

A span is an inline element that functions in the same ways as div (see Listing 4-4). Listing 4-5 constructs a span.

Listing 4-5. Div

```
<!DOCTYPE html>
<html>
<span> </span>
</html>
```

Buttons

A button enables users to click through and perform some action. Listing 4-6 constructs a button (see Figure 4-9).

Listing 4-6. HTML Button

```
<!DOCTYPE html>
<html>
<button type="button">Send</button>
</html>
```

Figure 4-9. *HTML button*

Text Box

A text box area enables users to insert text. Listing 4-7 constructs a text area input (see Figure 4-10)

Listing 4-7. HTML Text Area

```
<!DOCTYPE html>
<html>
<label for="Description">Description:</label>
<br>
<textarea id="textbox" name="textbox" rows="5" cols="60">
</textarea>
</html>
```

Figure 4-10. *HTML text area*

Input

At most, you would like users to input some data, like text or numeric data. Listing 4-8 constructs a numeric input (see Figure 4-11). `min` specifies the minimum numeric input, and `max` specifies the maximum numeric input.

Listing 4-8. HTML Numeric Input

```
<!DOCTYPE html>
<html>
<label for="Telephone">Telephone:</label>
<br>
<input type="number" type="number" min="1" max="20" >
</html>
```

Telephone:

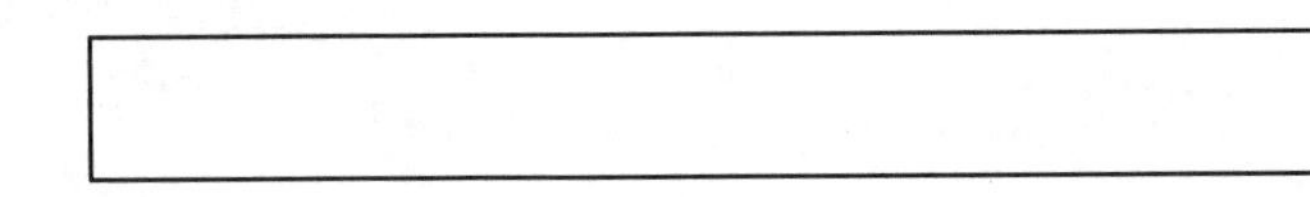

Figure 4-11. *HTML numeric input*

Listing 4-9 constructs a numeric input (see Figure 4-12).

Listing 4-9. HTML Single Line Text Input

```
<!DOCTYPE html>
<html>
<label for="FullName">Full Name:</label>
<br>
<input type="text" id="fullname" name=" fullname ">
</html>
```

Full Name:

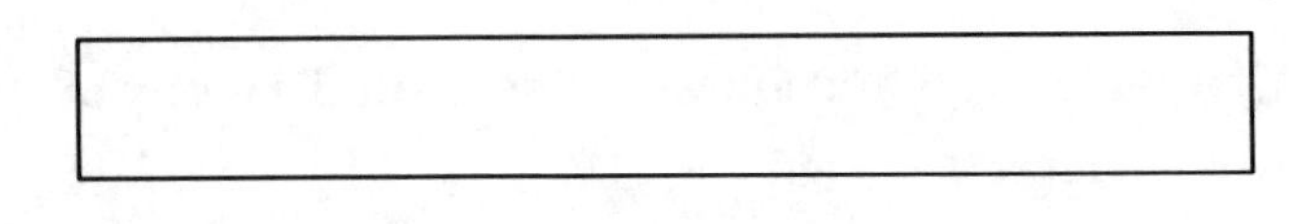

Figure 4-12. *HTML numeric input*

File Upload

You may want users to upload files. Listing 4-10 constructs a file upload input (see Figure 4-13).

Listing 4-10. HTML File Upload

```
<!DOCTYPE html>
<html>
<head>
<body>
<p> "Choose File" </p>
<form>
```

```
<input type="file" id="fileupload" name="fileupload">
</form>
</body>
</html>
```

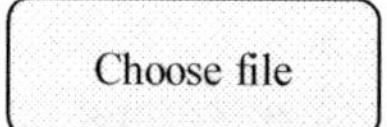

Figure 4-13. *HTML file upload*

Label

The label names an element. Listing 4-11 constructs a label.

Listing 4-11. HTML Label

```
</label>Full name</label>
```

Form

A form holds a group of elements and allows you to control a group of elements. Listing 4-12 constructs a form (see Figure 4-14).

Listing 4-12. Form

```
<!DOCTYPE html>
<html>
<head>
<body>
<p> "Choose File" </p>
<form>
<input type="file" id="fileupload" name="fileupload">
</form>
</body>
</html>
```

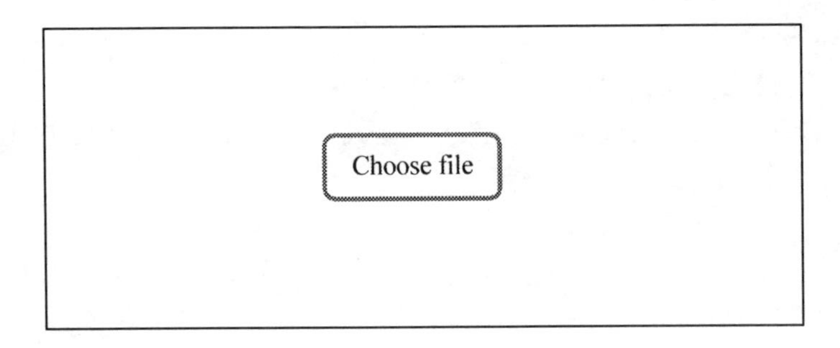

Figure 4-14. *HTML file upload*

Meta Tag

Meta tagging involves providing metadata (basic information) relating to a website (i.e., character set, keywords, the size of the web page, and scaling ratio, among others). Listing 4-13 constructs a meta tag.

Listing 4-13. HTML Meta Tag

```
<head>
  <meta charset="UTF-8">
  <meta name="description" content="Apress">
  <meta name="keywords" content="Programming, Technical, Machine Learning">
  <meta name="author" content="Apress Inc.">
  <meta name="viewport" content="width=device-width, initial-scale=1.0">
</head>
```

Practical Example

To put everything into perspective, Listing 4-14 presents a form that requires a user to input their basic information (i.e., full name, surname, telephone, upload a file, enter a comment) and submit it (see Figure 4-15).

Listing 4-14. HTML Form Example

```
<!DOCTYPE html>
<html>
<head>
  <meta charset="UTF-8">
  <meta name="description" content="Apress">
```

```
  <meta name="keywords" content="Programming, Technical, Machine Learning">
  <meta name="author" content="Apre">
  <meta name="viewport" content="width=device-width, initial-scale=1.0">
</head>
<form>
<h1>Submit a book to Apress</h1>
<label for="FullName">Full Name:</label>
<br>
<input type="text" name="FullName">
<br>
<label for=" Surname">Surname:</label>
<br>
<input type="text" name="Surname">
<br>
<label for="Telephone">Telephone:</label>
<br>
<input type="number" type="number" min="1" max="20" >
<br>
<label for="country">Country:</label>
<br>
<select name="country" id="country"
<option value="India">India</option>
<option value="SouthAfrica">South Africa</option>
</select>
<br>
<label for="fileselect">Upload Book</label>
<br>
       <input type="file" name="upload" id="fileselect">
<br>
<label for="Description">Message:</label>
<br>
<textarea rows="5" cols="50" name="Message" id="Description"></textarea>
<br>
<input type="submit" value="Submit">
</form>
</html>
```

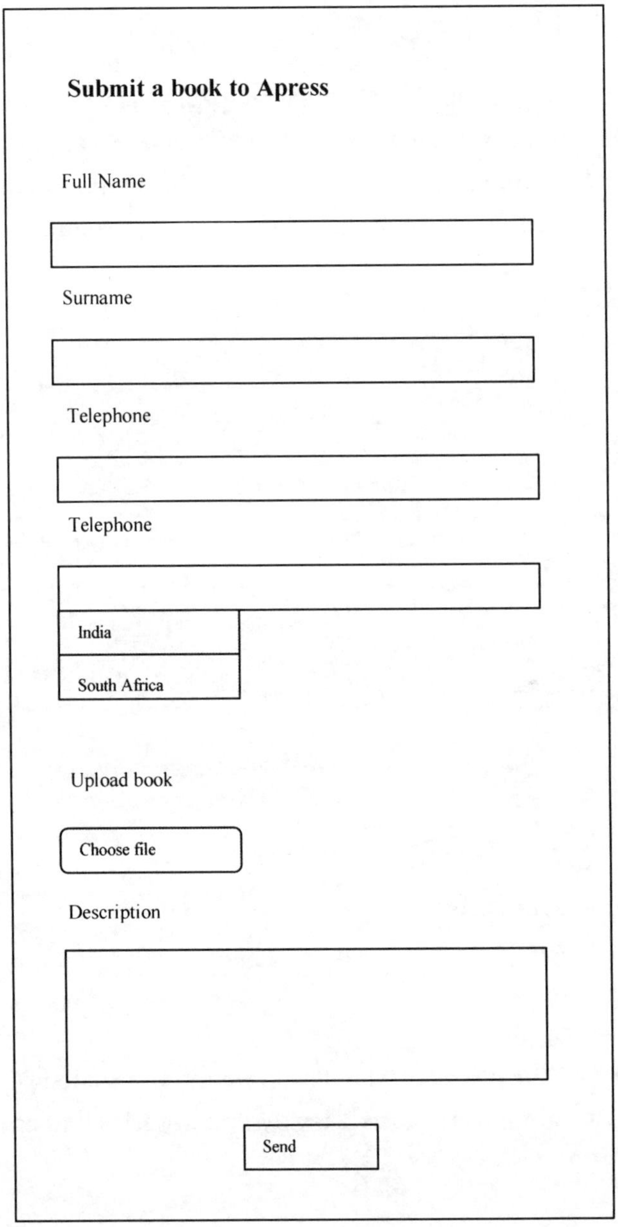

Figure 4-15. *HTML form example*

Viewing Web Page Source

You may view the source code of any web page by specifying `view-source:` before the web page URL; for example, see the `view-source:` at `www.apress.com/gp/book/9781484271094` (see the output in Figure 4-16). This helps you understand the basic structure of the HTML code, including web technologies implemented to build a web page.

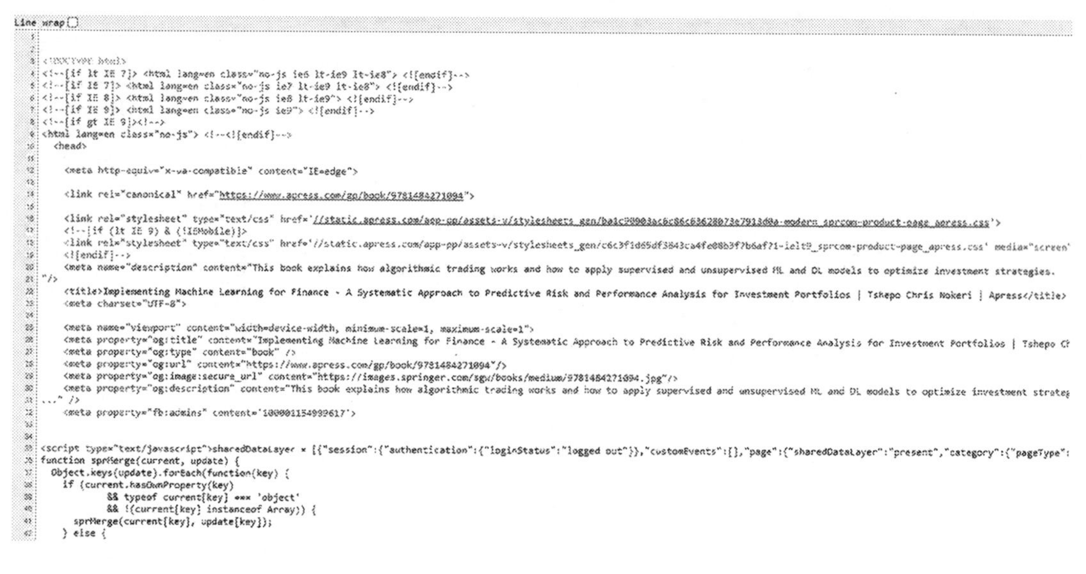

Figure 4-16. *View source code*

Conclusion

This chapter introduced the basics of web development. It revealed key HTML elements with code examples. Chapter 5 introduces key Python libraries that are useful for web development.

CHAPTER 5

Python Web Frameworks and Apps

Chapter 4 introduced you to interactive visualization using Plotly. This chapter introduces key Python web frameworks (i.e., Flask and Dash) and how they differ. Besides that, it provides practical examples that will help you get started with Python web app development.

First, you must install Flask and Dash. To install Flask in a Python environment, use `pip install flask`. Likewise, in a conda environment, use `conda install -c anaconda flask`.

To install Dash in a Python environment, use `pip install dash`. In a conda environment, use `conda install -c conda-forge dash`.

You must also install Dash HTML Components. In a Python environment, use `pip install dash-html-components`. In a conda environment, use `conda install -c conda-forge dash-html-components`.

Web Frameworks

A web framework typically makes up a chunk of code to design and deploy web apps without extensive programming. It purportedly contains code that runs in the background, thus avoiding your reinventing the wheel. There are many Python libraries (i.e., Flask and Dash). Their use depends on the context. This book typically implements a Dash library to design and deploy web apps. Besides what this book covers, there are other useful frameworks (e.g., Streamline and Bokeh).

T. C. Nokeri, *Web App Development and Real-Time Web Analytics with Python*,
https://doi.org/10.1007/978-1-4842-7783-6_5

Web Apps

The term *web app* is self-explanatory. It is essentially an app that runs on the web. Unlike traditional apps, you can access web apps anywhere by using any device that meaningfully connects to the web through a web browser. This sufficiently reveals a holistic approach to employing the Python programming language to develop web apps that contain interactive graphs and machine learning models.

Flask

Flask the most prevalent Python web framework. It is easy to use and has functionalities and elements that let you build upon HTML, JavaScript, and other languages.

WSGI

Flask implements the most prevalent Python standard for establishing a connection between a web browser and web server—the Web Server Gateway Interface (WSGI). WSGI has two sides: server/gateway and app/framework.

The server/gateway side makes up an end that establishes a connection with an Apache or Nginx web server. Apache is an open-source platform that provides many frameworks for big data processing, streaming, and machine learning. Nginx is another prevalent web server. Besides providing a gateway for web routing, it performs caching (temporarily storing events input, media, and files for reuse), load balancing (an act of stabilizing the traffic load), and reserve proxy (an act of validating the visible presence of information from the cache).

The app/framework side constitutes an end to that Python web app or framework.

Werkzeug

Werkzeug is grounded as WSGI, entialing that is based on the server/gateway and app/ framework as mentioned in the section overhead. It is an engine that selects templates that request and respond, among other tasks.

Jinja

Jinja is another template engine for Flask web apps. You should have it in your environment. To install Jinja, use `pip install jinja`. Listing 5-1 exhibits the essential Jinja structure by implementing the `template()` method.

Listing 5-1. Jinja Basic Structure

```
from jinja2 import Template
name = "Tshepo"
middle_name = "Chris"

tm = Template("My name is {{ name }} and my second name is {{ middle_name }}")
names = tm.render(name=name, middle_name = middle_name)
print(names)
```

This book does not implement Jinja, since jinja is based on the Flask framework, thus covering Flask will give you an understanding of it. There is no need to gain extensive knowledge about the library. Learn more about Jinja at `https://jinja.palletsprojects.com/en/3.0.x/`.

Installing Flask

First, import the Flask library and its key dependencies, as shown in Listing 5-2.

Listing 5-2. Importing Flask

```
from flask import Flask
```

Initializing a Flask Web App

Before running a Flask app, you must initialize it, as shown in Listing 5-3.

Listing 5-3. Initializing a Flask Web App

```
app = Flask(__name__)
```

Flask App Code

After importing Flask dependents, write the app code. Listing 5-4 writes Python code for a Flask app.

Listing 5-4. Flask App Code

```
@app.route("/")
def index():
    return "An Apress book"
```

Deploy a Flask Web App

After coding the structure of a Flask web app, you deploy the app on the localhost or cloud host. The code varies depending on the host. Listing 5-5 enables deploying a Flask web app on the localhost. Setting the mode to `offline` enables you to deploy a Flask web app offline.

Listing 5-5. Deploy a Flask Web App on Local Host

```
if __name__ == '__main__':
   app.run_server(debug=True)
```

The port number for a Flask app is 8050 (`https://localhost:8050`) if the intention is to deploy the app on an external host and specify the host ID.

Listing 5-6 deploys a Flask app to a specific host.

Listing 5-6. Deploy a Flask Web App on a Specific Host

```
if __name__ == '__main__':
   app.run_server(host='127.0.0.1', port=8050)
```

Dash

Dash is a web framework from the Plotly family. It is a prevalent Python web framework for designing and deploying dashboards. It works similarly to Flask, but it is simpler than the Flask library. This book implements the Dash library alongside Plotly.

Installing Dash Dependencies

After importing Flask dependents, write the app code. Listing 5-7 writes Python code for a Flask app.

Listing 5-7. Initializing a Dash app

```
import dash
```

Initializing a Dash Web App

Before running a Dash app, you must initialize it, as shown in Listing 5-8.

Listing 5-8. Initializing a Dash Web App

```
import dash
app = dash.Dash(__name__)
```

Dash Web App Code

After importing Dash dependents, write the app code. Listing 5-9 writes Python code for a Dash app.

Listing 5-9. Dash App Code

```
import dash
import dash_html_components as html
from jupyter_dash import JupyterDash
from dash.dependencies import Input, Output, State
app = JupyterDash(__name__)
app.layout = html.Div(
    [
        html.H2(id = 'book-output',
                children = ''),
    html.Button('Click to here learn about what this book is about',
                id='button')
    ],
    className = 'container')
```

```
@app.callback(
    Output('book-output', 'children'),
    [Input('button', 'n_clicks')])
def this_is_about(n_clicks):
    if n_clicks:
        return "This a book about web app development with Python"
if __name__ == '__main__':
    app.run_server(debug = True)
```

Deploy a Dash Web App

After coding the structure of a Dash web app, you deploy the app on the localhost or cloud host. The code varies depending on the host. Listing 5-10 enables a Dash app on the localhost.

Listing 5-10. Deploy a Dash Web App on Local Host

```
import dash
app = dash.Dash(__name__)
if __name__ == '__main__':
   app.run_server(debug=True)
```

Let's assume you want to deploy the app on an external host and specify the host ID. Listing 5-11 deploys a Dash app to a specific host.

Listing 5-11. Deploy a Dash Web App on a Specific Host

```
if __name__ == '__main__':
   app.run_server(host='127.0.0.1', port=8050)
```

Jupyter Dash

Jupyter Notebook is the most prevalent inline notebook for Python projects. It comes in handy with the Anaconda distribution platform. Cloud platforms like IBM Cloud, Amazon Web Services, and Microsoft Azure also use Jupyter Notebook.

Jupyter Dash is a Python library that enables you to design Dash apps from a Jupyter Notebook or JupyterLab. First, install Jupyter Dash in your Python environment using `pip install jupyter-dash`. Then, to get it running, use the following command: `"jupyter lab build"`.

Listing 5-12 imports Jupyter Dash, initializes a Dash web app, and writes code. Afterward, it specifies the layout. It concludes by deploying a web app.

Listing 5-12. Deploy a Dash Web App on a Specific Host

```
from jupyterdash import JupyterDash
app = JupyterDash(__name__)
app.layout = html.Div()
if _name_ == '__main__':
    app.run_server(mode="jupyterlab")
```

By default, the Dash library runs the app in debug mode, meaning that it debugs the code when you run it. While still in development, specify `dev_tools_ui` as `False` and `dev_tools_props_check` as `False` (see Listing 5-13).

Listing 5-13. Deploy a Dash Web App on a Specific Host

```
app.run_server(mode = "external",
               dev_tools_ui = False,
               dev_tools_props_check = False)
```

Conclusion

This chapter familiarizes you with two Python web frameworks—Flask and Dash. The subsequent chapters implement the Plotly library alongside the Dash library to create interactive dashboards for web apps. Make sure that you understand Dash in this chapter before proceeding to the next chapters.

CHAPTER 6

Dash Bootstrap Components

This chapter covers Dash Bootstrap Components, a Python library from the Plotly family that enables key Bootstrap functionalities on a Dash web app, thus simplifying web app development. Bootstrap borrows from HTML, CSS, and JavaScript. Figure 6-1 illustrates the building blocks of Bootstrap, thus `dash_bootstrap_components`.

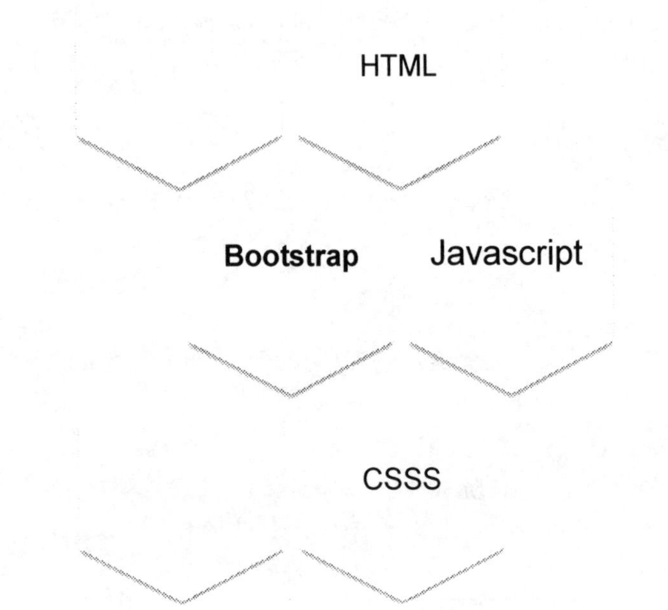

Figure 6-1. *Bootstrap*

After reading the content of this chapter, you should be able to implement key `dash_bootstrap_components` methods.

T. C. Nokeri, *Web App Development and Real-Time Web Analytics with Python*,
https://doi.org/10.1007/978-1-4842-7783-6_6

Dash Bootstrap Components. Install it in a Python environment using `pip install dash-bootstrap-components`. Likewise, install it in a conda environment using `conda install -c conda-forge dash-bootstrap-components`.

Listing 6-1. Dash Bootstrap Components

```
import dash
import dash_html_components as html
import dash_bootstrap_components as dbc
import dash_core_components as dcc
```

Number Input

Listing 6-2 constructs a numeric input by implementing the `Input()` method from the `dash_core_components` library (see Figure 6-2).

Listing 6-2. Number Input

```
app = JupyterDash(external_stylesheets=[dbc.themes.BOOTSTRAP])
number_input = html.Div(
    [
        html.P("Age"),
        dbc.Input(type="number", min=0, max=65, step=1),
    ],
    id="styled-numeric-input",
)
app.layout = html.Div([dcc.Location(id="url"), number_input])
app.run_server(mode = "external",
               dev_tools_ui = False,
               dev_tools_props_check = False)
```

Age

Figure 6-2. *Numeric input*

Text Area

Listing 6-3 constructs a text area by implementing the `Textarea()` method from the `dash_core_components` library (see Figure 6-3).

Listing 6-3. Text Area

```
app = JupyterDash(external_stylesheets=[dbc.themes.BOOTSTRAP])
textareas = html.Div(
    [
        dbc.Label("Comment"),
        html.Br(),
        dbc.Textarea(
            bs_size="lg",
            placeholder="Enter comment"
        ),
    ]
)
app.layout = html.Div([dcc.Location(id="url"), textareas])
app.run_server(mode = "external",
               dev_tools_ui = False,
               dev_tools_props_check = False)
```

Comment

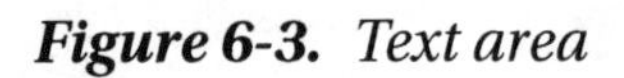

Figure 6-3. *Text area*

Select

Listing 6-4 constructs selection by implementing the `Select()` method from the `dash_bootstrap_components` library (see Figure 6-4).

Listing 6-4. Select

```
app = JupyterDash(external_stylesheets=[dbc.themes.BOOTSTRAP])
select = html.Div([
    dbc.Label("Select Gender")
    dbc.Select(
    id="select",
    options=[
        {"label": "Male", "value": "1"},
        {"label": "Female", "value": "2"}
    ],
)
])
app.layout = html.Div([dcc.Location(id="url"), select])
app.run_server(mode = "external",
               dev_tools_ui = False,
               dev_tools_props_check = False)
```

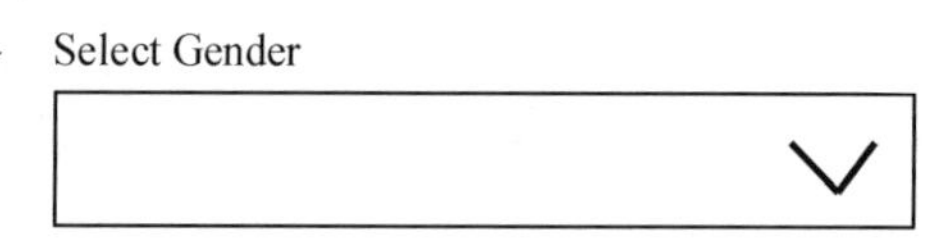

Figure 6-4. *Select*

Radio Items

Listing 6-5 constructs radio items by implementing the `RadioItems()` method from the `dash_bootstrap_components` library (see Figure 6-5).

Listing 6-5. Radio Items

```
app = JupyterDash(external_stylesheets=[dbc.themes.BOOTSTRAP])
radioitems = dbc.FormGroup(
    [
        dbc.Label("Choose preferred programming language"),
        dbc.RadioItems(
            options=[
```

```
                {"label": "Python", "value": 1},
                {"label": "R", "value": 2},
            ],
            value=1,
            id="radioitems-input",
        ),
    ]
)
app.layout = html.Div([dcc.Location(id="url"), radioitems])
app.run_server(mode = "external",
               dev_tools_ui = False,
               dev_tools_props_check = False)
```

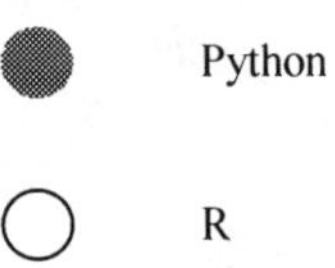

Figure 6-5. *Radio item*

Checklist

Listing 6-6 constructs a checklist by implementing the `Checklist()` method from the `dash_bootstrap_components` library (see Figure 6-6).

Listing 6-6. Checklist

```
app = JupyterDash(external_stylesheets=[dbc.themes.BOOTSTRAP])
checklist = dbc.FormGroup(
    [
        dbc.Label("Check web technologies you have experience with"),
        dbc.Checklist(
            options=[
                {"label": "HTML", "value": 1},
                {"label": "JavaScript", "value": 2},
                {"label": "CSS", "value": 3}
```

```
            ],
            value=[1],
            id="checklist-input",
        ),
    ]
)
app.layout = html.Div([dcc.Location(id="url"), checklist])
app.run_server(mode = "external",
               dev_tools_ui = False,
               dev_tools_props_check = False)
```

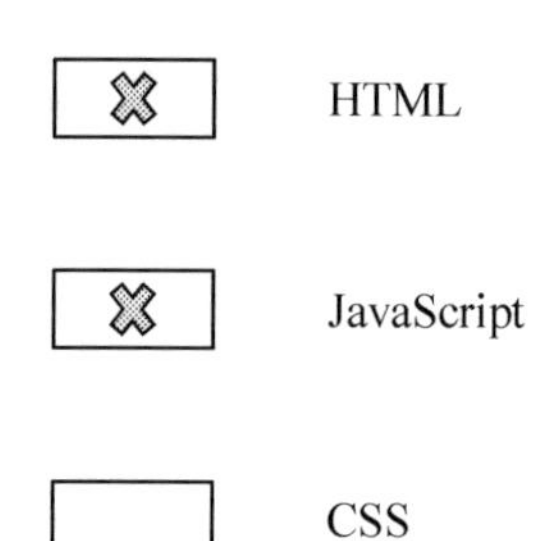

Figure 6-6. *Checklist*

Switches

Listing 6-7 constructs switches by implementing the `Checklist()` method from the `dash_bootstrap_components` library (see Figure 6-7).

Listing 6-7. Switches

```
app = JupyterDash(external_stylesheets=[dbc.themes.BOOTSTRAP])
switches = dbc.FormGroup(
    [
        dbc.Label("Do you enjoy reading programming books? (No/Yes)?"),
        dbc.Checklist(
            options=[
                {"label": "Yes", "value": 1},
            ],
```

```
            value=[1],
            id="switches-input",
            switch=True,
        ),
    ]
)
app.layout = html.Div([dcc.Location(id="url"), switches])
app.run_server(mode = "external",
               dev_tools_ui = False,
               dev_tools_props_check = False)
```

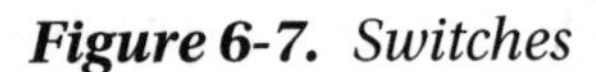

Figure 6-7. *Switches*

Tabs

Listing 6-8 constructs tabs by implementing the `Tab()` method from the `dash_bootstrap_components` library (see Figure 6-8).

Listing 6-8. Tabs

```
app = JupyterDash(external_stylesheets=[dbc.themes.BOOTSTRAP])
tab1_content = html.P("Tab 1 content")
tab2_content = html.P("Tab 2 content")
tabs = dbc.Tabs(
    [
        dbc.Tab(tab1_content, label="Tab 1"),
        dbc.Tab(tab2_content, label="Tab 2")
    ]
)
```

```
app.layout = html.Div([dcc.Location(id="url"), tabs])
app.run_server(mode = "external",
               dev_tools_ui = False,
               dev_tools_props_check = False)
```

Figure 6-8. *Switches*

Button

Listing 6-9 constructs a button by implementing the `Button()` method from the `dash_bootstrap_components` library (see Figure 6-9).

Listing 6-9. Button

```
app = JupyterDash(external_stylesheets=[dbc.themes.BOOTSTRAP])
callback_button = html.Div(
    [
        dbc.Button("Submit",
                   id = "submit-button",
                   className = "mr-2",
                   n_clicks = 0,
```

```
                    color = "light"
        )
    ]
)

app.layout = html.Div([dcc.Location(id="url"), callback_button])
app.run_server(mode = "external",
               dev_tools_ui = False,
               dev_tools_props_check = False)
```

Submit

Figure 6-9. *Switches*

Table

Apply dash_bootstrap_components alongside dash_html_components to create an HTML table from scratch. Listing 6-10 creates an HTML table. Tr represents a row, Th represents a column, and Td represents the input in a cell (see Table 6-1).

Listing 6-10. Constructing an HTML Table

```
app = JupyterDash(external_stylesheets=[dbc.themes.BOOTSTRAP])
table_header = [
    html.Thead(html.Tr([html.Th("Country"), html.Th("GNI")]))]
row1 = html.Tr([html.Td("Kazakhstan"), html.Td("8680")])
row2 = html.Tr([html.Td("Kenya"), html.Td("1760")])
row3 = html.Tr([html.Td("Kyrgyz Republic"), html.Td("1160")])
row4 = html.Tr([html.Td("Cambodia"), html.Td("1490")])
table_body = [html.Tbody([row1, row2, row3, row4])]
table = dbc.Table(table_header + table_body, bordered=True, hover=True,
responsive=True,striped=True)
app.layout = html.Div([dcc.Location(id="url"), table])
app.run_server(mode = "external",
               dev_tools_ui = False,
               dev_tools_props_check = False)
```

***Table 6-1.** HTML Table*

Country	GNI
Kazakhstan	8680
Kenya	1760
Kyrgyz Republic	1160
Cambodia	1490

Alternatively, convert a pandas DataFrame into an HTML table component by implementing the `from_dataframe` method from the `dash_bootstrap_components` library (see Listing 6-11 and Table 6-2).

***Listing 6-11.** Constructing an HTML Table*

```
import pandas as pd
app = JupyterDash(external_stylesheets=[dbc.themes.BOOTSTRAP])
df = pd.read_csv(r"filepath\data.csv")
table = dbc.Table.from_dataframe(df, striped=True, bordered=True,
hover=True)
app.layout = html.Div([dcc.Location(id="url"), table])
app.run_server(mode = "external",
               dev_tools_ui = False,
               dev_tools_props_check = False)
```

***Table 6-2.** Dataframe*

	gdp_by_exp	cpi	m3	spot_crude_oil	rand
DATE					
2009-01-01	-1.718249	71.178127	13.831098	41.74	9.3000
2009-04-01	-2.801610	73.249160	9.774203	49.79	9.3705
2009-07-01	-2.963243	74.448179	5.931918	64.09	7.7356
2009-10-01	-2.881582	74.884186	3.194678	75.82	7.7040
2010-01-01	0.286515	75.320193	0.961220	78.22	7.3613

Conclusion

This chapter covered key Dash Bootstrap components useful to building functional web apps. There are other components that you may include in the application. Learn more at the official Dash Bootstrap Components website (`https://dash-bootstrap-components.opensource.faculty.ai/`). Chapter 7 introduces implementing styles and themes to universally set the look and feel of components.

CHAPTER 7

Styling and Theming a Web App

This chapter introduces the basics of styling and theming a web app. First, it introduces styling an HTML web page. Subsequently, it acquaints you with the Cascade Styling Sheet (CSS) and calls a CSS code inside the architecture of a Dash web app. Following that, it presents the Bootstrap technology, including an approach to universally set the theme of a Dash web app by employing default theme templates, including those from an external source. Following that, it familiarizes you with a technique of setting the layout of a web app using the `dash_bootstrap_components` library.

Styling

Styling enables you to customize the look and feel of an HTML page. For instance, you can specify the border, background color, text color, transitions, and essential responses. Chapter 3 introduced the `div` element.

Listing 7-1 exhibits a way to create `div`. Notice that it applies `<style> </style>` to set `div` with `5px outset gray`, a `background-color` that is `white` and `text-align` is `center`.

T. C. Nokeri, *Web App Development and Real-Time Web Analytics with Python*,
https://doi.org/10.1007/978-1-4842-7783-6_7

Listing 7-1. Div

```
<html>
<head>
<style>
.div_1{
  border: 5px outset gray;
  background-color: white;
  text-align: center;
}
</style>
</head>
<body>

<div class="div_1">
  <h2>Apress Books</h2>
  <p>Books brought to you by Apress</p>
</div>

</body>
</html>
```

Cascade Styling Sheet

CSS is the most prevalent styling technique. It is implemented alongside HTML to control the design of a page. You apply CSS for styling by containing the code inside an HTML page or use an external CSS file.

Listing 7-2 exhibits an approach to contain CSS code inside HTML code.

Listing 7-2. CSS

```
<!DOCTYPE html>
<html>
<head>
<style>
body {background-color: gray;}
h1   {color: navy;}
```

```
p    {color: orange;}
</style>
</head>
<body>
<h1>An Apress Book</h1>
<p> This book introduces you to web application development and deployment
using Python web frameworks >
</body>
</html>
```

Listing 7-3 references an external CSS file.

Listing 7-3. Referencing a CSS File

```
<!DOCTYPE html>
<html>
<head>
  <link rel="stylesheet" href="styles.css">
</head>
<body>

<h1>An Apress Book</h1>
<p>This book introduces you to web application development and deployment
using Python web frameworks>

</body>
</html>
```

Listing 7-4 contains CSS code specifications. As you see, most of the arguments are self-explanatory. rem specifies the size. If you are familiar with pixels (`px`), you may specify the size as such. Pixels represent manageable elements populating the screen of any digital device. You can control the output on the screen by manipulating the properties. For instance, specifying `10px` display the content larger than specifying `5px`. Specifying `10rem` displays the content larger than specifying `5rem`.

Listing 7-4. CSS Code

```
css_style = {"position": "fixed",
             "top": 0,
             "left": 0,
             "bottom": 0,
             "width": "14rem",
             "height": "100%",
             "margin-bottom": "0rem",
             "padding": "0.5rem 1rem",
             "background-color": "#f8f9fa"}
```

If you want to implement the CSS code in Listing 7-4 to some component, specify it as the style (see Listing 7-5).

Listing 7-5. Calling Specific CSS

```
html = Div([], style=css_style)
```

Bootstrap

Bootstrap is a prevalent library for theming web applications. It integrates HTML, CSS, and JavaScript. There is no need for you to have extensive CSS programming knowledge or experience. It is an open source library initially developed by Twitter. It enables the design of responsive grid systems and offers a myriad of components, including some functionalities.

A key feature of the innovative Bootstrap framework enables the specific content of a web app to scale irrespective of the device. It is constructed based on putting mobile devices first. It is immensely impressive since you do not have to typically rewrite the coding for mobile devices.

An alternative web technology to Bootstrap is Google's Material Design. It was developed for structuring user interaction, and more specifically, screen orientation. The main differentiator between it and Bootstrap is that Material Design inclines more toward the appearance of a web app. Meanwhile, Bootstrap takes a remote approach that incorporates components and behavioral control, including appearance controls. Given this, Bootstrap covers a wider spectrum of web development than Material

Design. Moreover, it is very straightforward to learn. You can learn more about it at https://material.io/design/introduction.

Listing 7-6 sources a CSS file from the prevalent Bootstrap theming library. Learn more at https://getbootstrap.com.

Listing 7-6. Sourcing External CSS File

```
getbootstrap = "https://cdn.jsdelivr.net/npm/bootstrap@5.0.2/
dist/css/bootstrap.min.css" rel="stylesheet" integrity="sha384-
EVSTQN3/azprG1Anm3QDgpJLIm9NaoOYz1ztcQTwFspd3yD65VohhpuuCOmLASjC"
crossorigin="anonymous"
app = dash.Dash(external_stylesheets=[getbootstrap])
```

If you want to employ an icon from an external library, specify its URL (see Listing 7-7). For this example, the icons are sourced from Font Awesome, a prevalent icon library.

Listing 7-7. Specify Theme Template and Icons Library

```
FA = https://use.fontawesome.com/releases/v5.8.1/css/all.css
app = JupyterDash(external_stylesheets=[dbc.themes.MATERIA, FA])
```

The best way to integrate icons into a web app is by specifying the icon's name as the className argument, which calls CSS functionalities without rewriting the code for each use. For instance, if you desire to employ the alert icon, specify the className as "far fa-bell". Likewise, to include the message icon, use "far fa-envelope". This applies to all the icons. Learn more about Font Awesome icons at https://fontawesome.com/v4.7/icons/.

Dash Bootstrapping

This book implements the dash_bootstrap_components library, which is based on Bootstrap technology. As a result, creating dashboards and web apps is not a painstaking experience. There is no need for you to have extensive Bootstrap programming knowledge or experience.

First, let's import Dash (see Listing 7-8).

Listing 7-8. Import Dash

```
import dash
```

Dash Core Components

The dash_core_components library implements key Dash functionalities. First, ensure that you have installed the dash_core_components library in your environment. To install it in a Python environment, use pip install dash-core-components. Likewise, to install it in a conda environment, use conda install -c conda-forge dash-core-components.

Listing 7-9 imports dash_core_components.

Listing 7-9. Import Dash Core Components

```
import dash_core_components as dcc
```

Dash Bootstrap Components

The dash_bootstrap_components library implements Bootstrap functionalities. First, ensure that you have installed the dash_bootstrap_components library in your environment. To install it in a Python environment, use pip install dash-bootstrap-components. To install it in a conda environment, use conda install -c conda-forge dash-bootstrap-components.

Listing 7-10 imports dash_bootstrap_components.

Listing 7-10. Dash Bootstrap Components

```
import dash_bootstrap_components as dbc
```

Implementing Dash Bootstrap Components Theming

Listing 7-11 obtains the theming scheme within Dash by specifying the external_stylesheets argument (for this example, specify it as BOOTSTRAP). Assuming there is a navigation bar, the outcome looks like Figure 7-1.

Listing 7-11. Implementing Dash Bootstrap Components Theming

```
app = dash.Dash(external_stylesheets=[dbc.themes.BOOTSTRAP])
```

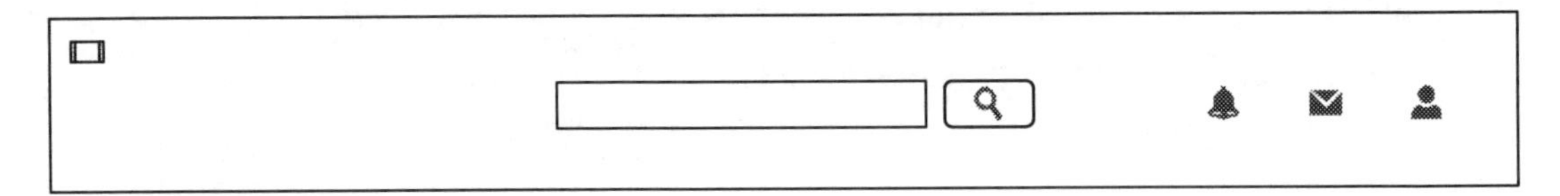

Figure 7-1. *Navigation bar*

In addition to BOOTSTRAP, there are alternative themes like CERULEAN, COSMO, CYBORG, DARKLY, FLATLY, JOURNAL, LITERA, LUMEN, LUX, MATERIA, MINTY, PULSE, SANDSTONE, SIMPLEX, SKETCHY, SLATE, SOLAR, SPACELAB, SUPERHERO, UNITED, YETI.

Applying the example in Figure 7-1. Assuming there is a navigation bar and the theme is specified as DARKLY (see Listing 7-12), the outcome looks like Figure 7-2.

Listing 7-12. Implementing Dash Bootstrap Components Theming

```
app = dash.Dash(external_stylesheets=[dbc.themes.DARKLY])
```

Figure 7-2. *Navigation bar*

Applying the example in Figure 7-1. Assuming there is a navigation bar and the theme is specified as CERULEAN (see Listing 7-13), the outcome looks like Figure 7-3.

Listing 7-13. Implementing Dash Bootstrap Components Theming

```
app = dash.Dash(external_stylesheets=[dbc.themes.CERULEAN])
```

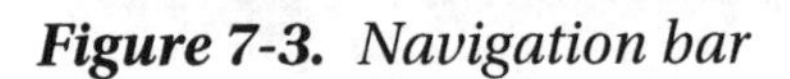

Figure 7-3. *Navigation bar*

Alternatively, extract a CSS file from an external website (see Listing 7-14).

Listing 7-14. Implementing Dash Bootstrap Components Theming

```
BS = "https://stackpath.bootstrapcdn.com/bootstrap/4.4.1/css/bootstrap.min.css"
app = dash.Dash(external_stylesheets=[BS])
```

Dash HTML Components

The `dash_html_components` library implements HTML functionalities. First, ensure that you have installed the `dash_html_components` library in your environment. To install it in a Python environment, use `pip install dash-html-components`. To install it in a conda environment, use `conda install -c conda-forge dash-html-components`.

Listing 7-15 imports `dash_html_components`.

Listing 7-15. Import Dash HTML Components

```
import dash_html_components as html
```

Dash Web Application Layout Design

Before running a Dash web application, you need to specify its layout structure. Listing 7-16 specifies the structure by implementing the `layout()` method.

Listing 7-16. Dash Web Application Layout Design

```
from jupyterdash import JupyterDash
app = JupyterDash(__name__)
@app.route("/")
def index():
    return "An Apress book"
app.layout = html.Div()
if __name__ == '__main__':
   app.run(mode='offline')
```

Responsive Grid System

A layout helps construct a responsive grid system. When doing so, ensure that you specify the number of rows and columns. A Bootstrap layout contains a row width that is 12 (see Figure 7-4).

Given the specifications, ensure that the width of each row does not exceed 12.

Listing 7-17. Grid System

```
row = html.Div([
    dbc.Row(
        dbc.Col([html.Div("Width = 12")],
                width=12)),
    dbc.Row([
        dbc.Col(html.Div("Width = 6"),
                width=6),
        dbc.Col(html.Div("Width = 6"),
                width=6)]),
    dbc.Row([
        dbc.Col(html.Div("Width = 4"),
                width=4),
        dbc.Col(html.Div("Width = 4"),
                        width=4),
        dbc.Col(html.Div("Width = 4"),
                        width = 4)])])
```

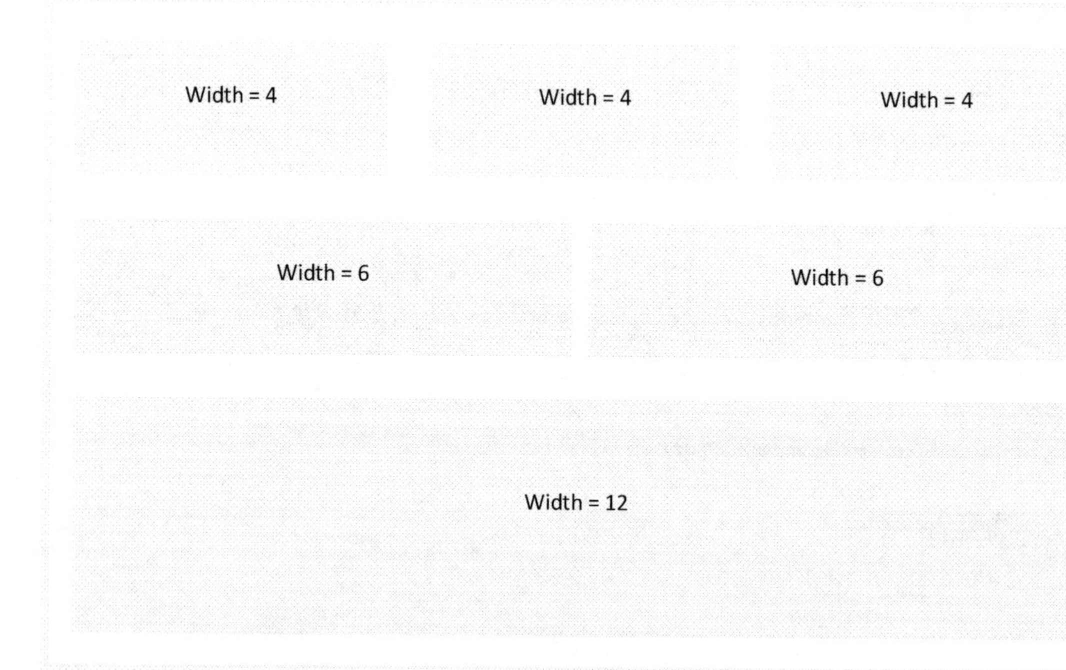

Figure 7-4. *Grid system*

Figure 7-4 presents a grid with three rows. The first row comprises three columns, the second row comprises two columns, and the third column comprises three columns.

Alternatively, you can specify each row separately (see Listing 7-17).

Listing 7-18. Grid System

```
row1 = dbc.Row([
    dbc.Col(html.Div("Width = 12"),
                        width=12)])
row2 = dbc.Row([
    dbc.Col(html.Div("Width = 6"),
            width=6),
    dbc.Col(html.Div("Width = 6"),
            width=6)])
row3 = dbc.Row([
    dbc.Col(html.Div("Width = 4"),
            width=4),
```

```
    dbc.Col(html.Div("Width = 4"),
            width=4),
    dbc.Col(html.Div("Width = 4"),
            width = 4)])
rows = html.Div([row1, row2, row3])
```

Conclusion

This chapter sufficiently acquaints you with the prime essentials of styling using CSS. Besides that, it promptly presented a holistic approach of universally setting the theme of a web app using some free Bootstrap theme templates that come alongside `dash_bootstrap_components`.

Chapter 8 focuses on developing real-time web analytics dashboards and web apps by integrating the Dash library, CSS, and Plotly with other standard Python libraries.

CHAPTER 8

Building a Real-Time Web App

This chapter promptly introduces you to ostensibly constructing real-time interactive web apps with Python, with a responsive navigation bar, sidebar, charts, tables, callbacks, and URL routing. After referring to the unique contents of this chapter, you should be skilled in developing a functional and responsive dashboard as web apps by implementing key web frameworks centered on HTML and Bootstrap technologies (i.e., Dash, Dash Core Components, Dash HTML Components, and Dash Bootstrap Components), alongside the Plotly interactive charting library. It sufficiently acquaints you with a holistic approach for implementing CSS to customize as a web app. And it includes functionalities like search, including report generation, and download functionality.

Approach this chapter using the following tips.

Tip Ensure that you assign an ID to each component so that you may reuse it.

Incorporate a CSS file or specify it in the Dash app to control the styling and behavior of components without rewriting the CSS code for each component. Alternatively, employ Bootstrap theme templates.

To install the dependencies required for this example, visit the source code folder, and use `pip install -r requirements.txt`. Also, refer to Listing 8-1.

T. C. Nokeri, *Web App Development and Real-Time Web Analytics with Python*,
https://doi.org/10.1007/978-1-4842-7783-6_8

Dash App Structure

Figure 8-1 illustrates an approach to structuring code for a Dash web app.

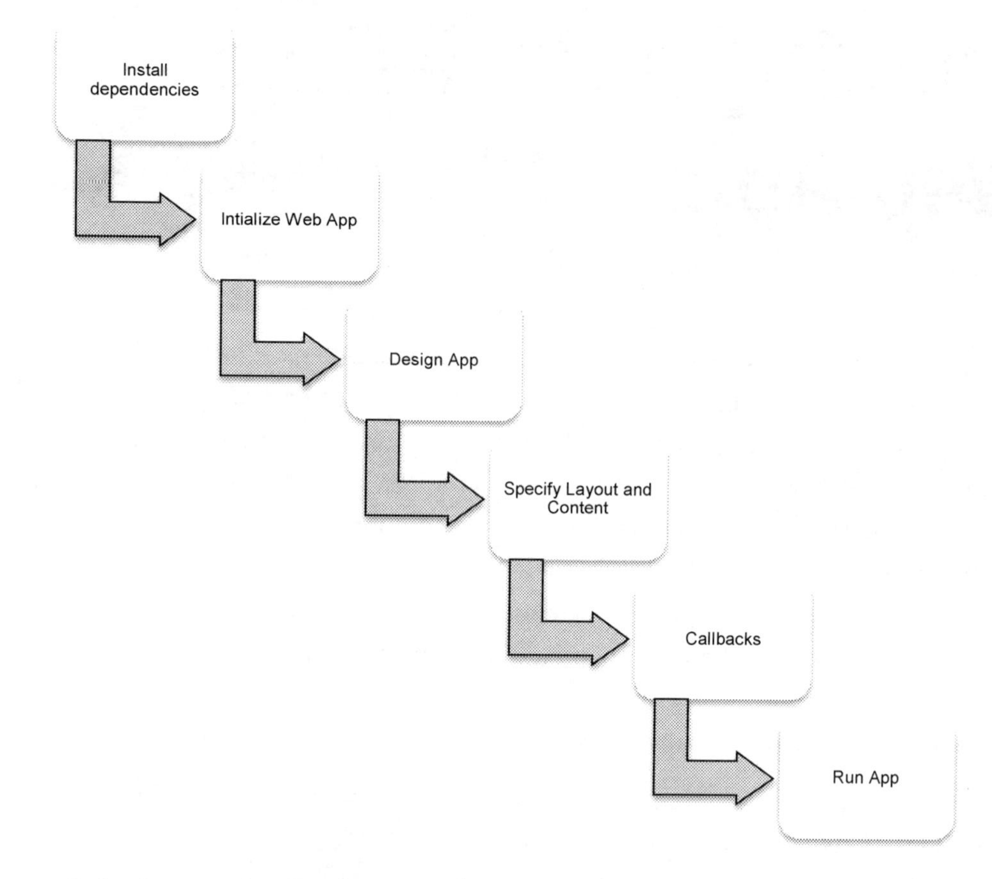

Figure 8-1. *Structuring Dash app code*

Figure 8-1 presents the steps as follows: import key dependencies, initialize the Dash web app, and specify the theme, construct components, specify the app layout and content, and then run the app.

Importing Key Dependencies

The initial step in building a Dash web app involves importing key dependencies.

Listing 8-1 imports key dependencies to design an optimal web app.

Listing 8-1. Importing Key Dependencies

```
import pandas as pd
import dash_table as dt
import plotly
import plotly.express as px
import plotly.graph_objects as go
import plotly.io as pio
pio.templates.default = "simple_white"
import dash
import dash_core_components as dcc
import dash_html_components as html
from dash.dependencies import Input, Output, State
import dash_bootstrap_components as dbc
from dash_extensions import Download
from dash_extensions.snippets import send_data_frame
from jupyter_dash import JupyterDash
from pandas_datareader import wb
import wbdata
import re
```

The next segment presents a strategy for incorporating search functionality into the app, the component enabling a user to search and select a country, and an indicator from a list of options derived from a Microsoft Excel file document.

Listing 8-2 extracts the data by implementing the `read_excel()` method from the pandas library. It captures the symbols and names of countries in the list to serve as `options` in the search drop-down menu in the Dash app by implementing the `append()` method from the pandas library.

Table 8-1 highlights the data contained in the Microsoft Excel file document.

Listing 8-2. Extracting Country Search Options

```
country_ticker = pd.read_excel(r"filepath\list_of_countries.xlsx")
country_ticker = country_ticker.set_index("Symbol")
country_options = []
```

```
for country_tic in country_ticker.index:
    country_options.append({"label":"{} {}".format(country_tic, country_
    ticker.loc[country_tic]["Name"]), "value":country_tic})
```

Table 8-1. *Country Listing*

	Unnamed: 0	Name
Symbol		
ABW	0	Aruba
AFG	1	Afghanistan
AFR	2	Africa
AGO	3	Angola
ALB	4	Albania
...	...	...
XZN	292	Sub-Saharan Africa excluding South Africa and ...
YEM	293	Yemen, Rep.
ZAF	294	South Africa
ZMB	295	Zambia
ZWE	296	Zimbabwe

Listing 8-3 extracts the data by implementing the `read_excel()` method from the pandas library. Subsequently, it captures the symbols and names of indicators in the list to serve as options in the search drop-down menu in the Dash app by implementing the `append()` method from the pandas library.

Table 8-2 highlights the data contained in the Microsoft Excel file document.

Listing 8-3. Extracting Indicators Search Options

```
global_ind_ticker = pd.read_excel(r"filepath\global_ind.xlsx")
global_ind_ticker = global_ind_ticker.set_index("Symbol")
global_ind_options = []
```

```
for global_ind_tic in global_ind_ticker.index:
    global_ind_options.append({"label":"{} {}".format(global_ind_tic,
    global_ind_ticker.loc[global_ind_tic]["Name"]), "value":global_ind_tic})
```

Table 8-2. *Indicator Listing*

	Name
Symbol	
AG.AGR.TRAC.NO	Agricultural machinery, tractors
AG.CON.FERT.PT.ZS	Fertilizer consumption (% of fertilizer produc...
AG.CON.FERT.ZS	Fertilizer consumption (kilograms per hectare ...
AG.LND.AGRI.K2	Agricultural land (sq. km)
AG.LND.AGRI.ZS	Agricultural land (% of land area)
...	...
VC.IDP.TOCV	Internally displaced persons, total displaced ...
VC.IHR.PSRC.FE.P5	Intentional homicides, female (per 100,000 fem...
VC.IHR.PSRC.MA.P5	Intentional homicides, male (per 100,000 male)
VC.IHR.PSRC.P5	Intentional homicides (per 100,000 people)
VC.PKP.TOTL.UN	Presence of peace keepers (number of troops, p...

Loading an External CSS File

Listing 8-4 extracts a CSS file from Bootstrap, a popular CSS provider (`https://getbootstrap.com/docs/5.1/getting-started/introduction/`). Alternatively, use another CSS provider, like Font Awesome (`https://use.fontawesome.com/releases/v5.8.1/css/all.css`).

Listing 8-4. Loading an External CSS File

```
get_bootstrap_css = "https://cdn.jsdelivr.net/npm/bootstrap@5.0.0-beta2/
dist/css/bootstrap.min.css"
```

Loading the Bootstrap Icons Library

To make a Dash web app more appealing, you can incorporate icons. Listing 8-5 extracts the Bootstrap Icons library. You can review it at `https://icons.getbootstrap.com`.

Listing 8-5. Loading the Bootstrap Icons Library

```
get_bootstrap_icon = "https://cdn.jsdelivr.net/npm/bootstrap-icons@1.4.0/
font/bootstrap-icons.css"
```

Initializing a Web App

After importing key Python libraries, loading the CSS file and icons library, the next step involves initializing the Dash web app. First, Listing 8-6 initializes the app by implementing the JupyterDash library. Following that, it specifies the `external_stylesheets` as the Bootstrap CSS file loaded in Listing 8-4 and icons library loaded in Listing 8-5.

It specifies `meta_tags` as “`name`”: “`viewport`”, “`charset`”:“`utf-8`”,“`content`”: “`width=device-width,initial-scale=1, shrink-to-fit=no`” to make the web app mobile-friendly.

Listing 8-6. Initializing a Web App

```
app = JupyterDash(external_stylesheets=[get_bootstrap_css, get_bootsrap_
icon],
                meta_tags=[{"name": "viewport",
                            "charset":"utf-8",
                            "content": "width=device-width,initial-
                            scale=1, shrink-to-fit=no"}])
```

Navigation Bars

Web apps comprise a navigation bar, which enables a user to scheme through the app. This chapter demonstrates examples of properly constructing a top navigation bar and side navigation bar. Initially, it constructs a top navigation bar with icons embedded in it.

Top Navigation Bar

The top navigation bar in this chapter comprises three navigation items: alerts and notifications (which routes a user to the "alerts and notification" page), message (routing a user to the inbox), and profile (releasing a drop-down menu comprising items like Edit Profile, Account Settings, Billing, and Sign Out; each item has its own link). Besides that, it comprises a toggle to hide and unhide the side navigation bar.

Alerts and Notifications

Listing 8-7 constructs alerts and notifications by implementing the `NavLink()` method from the `dash_bootstrap_components` library. It specifies the `className` as "`bi bi-alarm`" to extract an icon from the Bootstrap Icons library, and then it specifies the `font-size` at "`20px`", `color` as "`gray`", and `width` as "`auto`" to enable autoscaling depending on a user's device (see Figure 8-2). Likewise, it specifies the URL as `page-1/1`. Notice that the ID is specified as "`submenu-1-collapse`", and the following navigation item is specified as "`submenu-2-collapse`". Specifying the ID enables you to URL route all components at once, rather than writing a separate line of code for each.

Listing 8-7. Alerts and Notifications

```
messages = dbc.Row(
    [
        dbc.Col(
            dbc.NavLink(className = "bi bi-envelope",
                        href = "/page-3/1",
                        style = {"font-size" : "20px", "color" : "gray"}),
            width="auto"
        ),
    ],
    no_gutters = True,
    className = "ml-auto flex-nowrap mt-3 mt-md-0",
    align = "right",
     style = {"font-size" : "16px"},
    id = "responsivemenu-4-collapse"
)
```

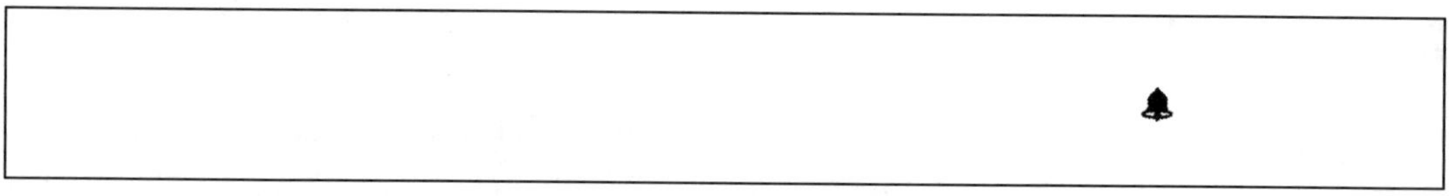

Figure 8-2. *Alerts and notifications*

Messages

Listing 8-8 constructs messages by implementing the NavLink() method from the dash_bootstrap_components library. First, it specifies the className as "bi bi-envelope" to extract an icon from the Bootstrap Icons library. Next, it specifies the font-size as "20px", color as "gray", and width as "auto" to enable autoscaling depending on a user's device (see Figure 8-3). In addition, it specifies the URL as page-1/2.

Listing 8-8. Messages

```
messages = dbc.Row(
    [
        dbc.Col(
            dbc.NavLink(className = "bi bi-envelope",
                        href = "/page-3/1",
                        style = {"font-size" : "20px", "color" : "gray"}),
            width="auto"
        ),
    ],
    no_gutters = True,
    className = "ml-auto flex-nowrap mt-3 mt-md-0",
    align = "right",
     style = {"font-size" : "16px"},
    id = "responsivemenu-4-collapse"
)
```

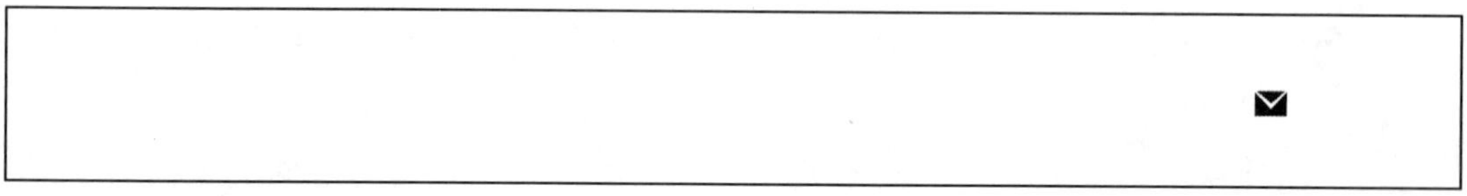

Figure 8-3. *Messages*

Profile

Listing 8-9 constructs a navigation item known as a *profile* by implementing the dbc.DropdownMenu() method from the dash_bootstrap_components library (see Figure 8-4). The drop-down menu comprises the following items: Edit Profile, Account Settings, Billing, and Sign Out, with each item comprising its own link (see href). The direction of the drop-down menu is "left".

Listing 8-9. Profile

```
profile = dbc.DropdownMenu(
    children=[
        dbc.DropdownMenuItem("Edit Profile",
                             href = "/page-4/1"),
        dbc.DropdownMenuItem("Privacy & Safety",
                             href = "/page-4/2"),
        dbc.DropdownMenuItem("Account Settings",
                             href = "/page-4/3"),
        dbc.DropdownMenuItem("Billing",
                             href = "/page-4/4"),
        dbc.DropdownMenuItem(divider = True),
        dbc.DropdownMenuItem("Sign Out",
                             href = "/page-4/5"),
    ],
    nav = True,
    in_navbar = True,
    className = "bi bi-person",
    direction = "left",
    style = {"font-size" : "20px"},
    id = "responsivemenu-5-collapse"
)
```

Figure 8-4. *Profile*

Navigation Bar with a Toggle Button

This section develops a navigation bar with a toggle button. A *toggle* collapses items (i.e., a responsive side navigation bar). It enables users to navigate an app through a mobile device easily, without displaying all components on the screen, since the screen width differs between laptops/PC monitors and mobile devices. In this chapter, it hides and unhides the responsive side navigation bar (see Figure 8-5).

Listing 8-10 completes the top navigation bar with a toggle button by implementing the `Navbar()` method from the `dash_bootstrap_components` library (see Figure 8-5). It specifies the `className` as "`navbar-toggler-icon`" and `n_clicks` (number of clicks) as `0`. Specifying `n_clicks` as `0` ensures that no action occurs before a user clicks the button. Besides that, it creates space between items in the drop-down by specifying "`padding`" as "`1rem 0rem`" and specifies the color of the navigation bar as "`white`".

Note There are empty columns for spacing. Bootstrap is built on responsive grid systems, which use columns (with a maximum width of 12) and rows to position the components in an app.

Listing 8-10. Navigation Bar

```
navbar = dbc.Navbar(
    [
        dbc.Col([],width=2),
        dbc.Col([
            dbc.Button(id = "toggle-button",
                       n_clicks = 0,
                       children = "",
                       outline = True,
                       className = "navbar-toggler-icon")],
            width = 1),
        dbc.Col([],
                width = 3),
        dbc.Col([messages],
                width = "auto"),
```

```
        dbc.Col([alerts_notif],
                width = "auto"),
        dbc.Col([], width = 2),
        dbc.Col([profile],
                width = 2)],
    color = "white",
    style={"margin-right" : "0rem",
           "margin-top" : "0rem",
           "margin-bottom" : "0.5rem",
           "padding" : "1rem 0rem"})
```

Figure 8-5. *Navigation bar*

Specifying the Responsive Side Navigation Bar

In addition to the top navigation bar, include a side navigation bar in the app, which comprises items not included in the top navigation bar. First, specify the CSS. Listing 8-11 specifies the CSS code for the side navigation bar. It also specifies the `background-color` as "`#f8f9fa`" (light gray), also it specifies `padding` (spacing between the elements) as "`0.5rem 1rem`".

Listing 8-11. Visible Sidebar Navigation Bar CSS Code

```
RESPONSIVE_MENU_STYLE = {
    "position" : "fixed",
    "top" : 0,
    "left" : 0,
    "bottom" : 0,
    "width" : "14rem",
    "height" : "100%",
    "margin-top" : "0rem",
    "margin-bottom" : "0rem",
    "z-index" : 1,
    "overflow-x" : "hidden",
    "transition" : "all 0.5s",
    "padding" : "0.5rem 1rem"
}
```

The best way to ensure that the sidebar navigation bar hides and unhides upon clicking the toggle button involves specifying a separate CSS code for the side. The callback section of this chapter ensures the side navigation bar hides and unhides upon clicking the toggle button.

Listing 8-12 specifies the CSS code that enables hiding and unhiding the sidebar navigation bar. The primary differentiator between the CSS code in Listing 8-11 and Listing 8-12 is that Listing 8-12 specifies "`left`" as "`-16rem`", which hides the sidebar navigation bar.

Listing 8-12. Hidden Sidebar Navigation Bar CSS Code

```
RESPONSIVE_MENU_HIDEN = {
    "position" : "fixed",
    "top" : 0,
    "left" : "-16rem",
    "bottom" : 0,
    "width" : "14rem",
    "height" : "100%",
    "z-index" : 1,
    "overflow-x" : "hidden",
```

```
    "transition" : "all 0.5s",
    "padding" : "0rem 0rem",
}
```

Specifying the Web App CSS Code

Importing crucial dependencies, initializing the app, and developing navigation bars subsequently develops the contents of the app. The content may contain interactive tables and charts, including components.

Listing 8-13 specifies the CSS code for the Dash app content.

Listing 8-13. Content CSS Code

```
APP_CONTENT_STYLE = {
    "transition" : "margin-left .5s",
    "margin-left" : "14.5rem",
    "margin-right" : "0.5rem",
    "margin-bottom" : "0.5rem",
    "padding" : "0rem 0rem"
}
```

This section develops a `toggle` that triggers the hiding or unhiding of the side navigation bar.

To stretch the app content when the side navigation bar is hidden, replace the "`margin-left`" from "`14rem`" to "`0rem`" (see Listing 8-14).

Listing 8-14. Content CSS Code

```
APP_CONTENT_STYLE1 = {
    "transition" : "margin-left .5s",
    "margin-left" : "0.5rem",
    "margin-right" : "0.5rem",
    "margin-bottom" : "0.5rem",
    "padding" : "0rem 0rem"
}
```

Side Navigation Bar Menus and Submenus

After specifying the CSS code for the sidebar navigation bar, the next step involves developing the items it must contain.

Figure 8-6 illustrates the structure of the sidebar navigation bar menu containing submenus.

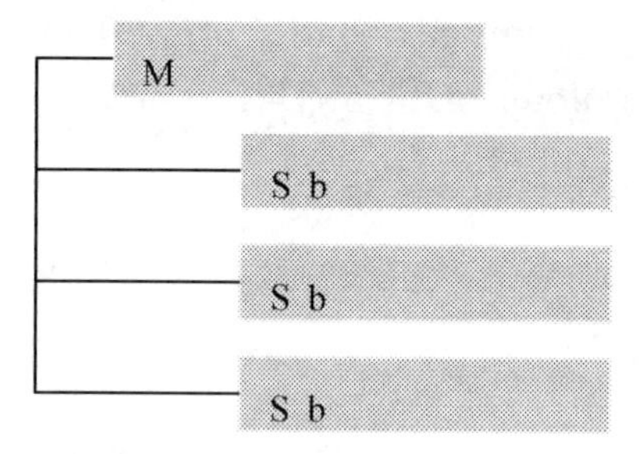

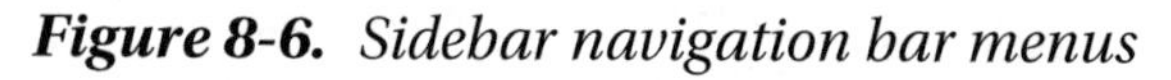

Figure 8-6. *Sidebar navigation bar menus*

Listing 8-15 develops menus and submenus by implementing the `NavLink()` method from the `dash_html_components` library.

Note that the ID for the Market menu is specified as "`responsivemenu-1`" and the Forecaster menu is "`responsivemenu-2`". Also, the ID for the submenu contained in the `Collapse()` method for "`Forecaster`" is "`responsivemenu-2-collapse`".

Listing 8-15. Developing Submenus

```
RESPONSIVE_RESPONSIVEMENU_1 = [
    html.Li(
        dbc.Row(
            [
                dbc.Col(dbc.NavLink("Search indicators",
                        href = "/page-1/1",
                        style = {"color" : "#616161"})),
            ],
            className = "my-1",
        ),
```

```
        style = {"cursor" : "pointer"},
        id = "responsivemenu-1",
    )
]
RESPONSIVE_RESPONSIVEMENU_2 = [
    html.Li(
        dbc.Row(
            [
                dbc.Col("Forecaster"),
                dbc.Col(
                    [html.I(className = "bi bi-chevron-down")],
                    width = "auto")],
            className = "my-1"),
        style = {"cursor" : "pointer"},
        id = "responsivemenu-2"
    ),
    dbc.Collapse(
        [
            dbc.NavLink("Forecast indicators",
                        href = "/page-2/1",
                        style = {"color" : "#616161"}),
        ],
        id = "responsivemenu-2-collapse"
    ),
]
```

After specifying the submenus, complete the side navigation bar.

Listing 8-16 completes the side navigation bar by implementing the `Nav()` from the `dash_bootstrap_components` library and containing them inside the `Card()` method (see Figure 8-7).

Note It specifies `vertical` as `True` to ensure the items are placed vertically.

Listing 8-16. Developing a Side Navigation Bar

```
RESPONSIVE_SIDE_NAVIGATION_BAR = html.Div(
    [
        dbc.Card(
            [
                dbc.CardBody(
                    [
                        html.H4("WorldViewer",
                                className = "btn btn-outline-primary"),
                        html.Hr(),
                        html.P(
                            "",
                            className = "lead"),
                        dbc.Nav(
                            RESPONSIVE_RESPONSIVEMENU_1,
                            vertical = True),
                        html.Hr(),
                        dbc.Nav(
                            RESPONSIVE_RESPONSIVEMENU_2,
                            vertical = True
                        ),
                    ]
                )
            ],
            id = "responsivesidebar")
    ]
)
```

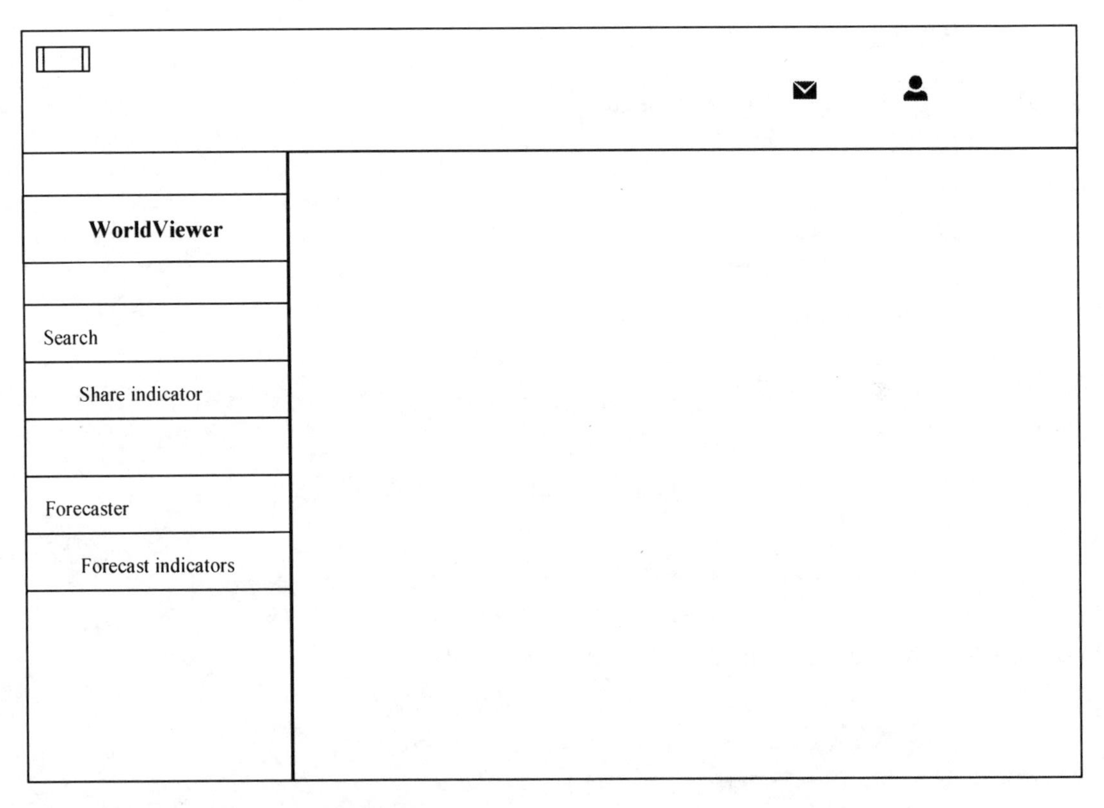

Figure 8-7. *Side navigation bar*

Search Functionality

Listing 8-17 develops a search component by implementing the `Dropdown()` method from the `dash_core_components` library. It specifies `options` as "`options`", which was specified in Listing 8-3. The purpose of this method is to avoid having to create a list of `options` independently.

It specifies `multi` as `False`, meaning the user can only select one country and indicator for each search procedure. Likewise, it specifies the placeholder (text that appears on the component informing a user of the type of input required) as "`Search country`", (`id` = "`country-symbol`") and "`Search indicator`" (`id` = "`country-indicator`").

Listing 8-17. Developing the Search Component

```
INPUT_CARD = dbc.Card([dbc.CardBody([
    dbc.Row([
        dbc.Col([
            dbc.Label("Select country"),
            html.Br(),
            dcc.Dropdown(id="country-symbol",
                         className="border-bottom",
                         options=country_options,
                         multi=False,
                         placeholder="Search country")]),
        dbc.Col([
            dbc.Label("Select indicator"),
            html.Br(),
            dcc.Dropdown(id = "country-indicator",
                         className=  "border-bottom",
                         options = global_ind_options,
                         multi = False,
                         placeholder = "Search indicator")])]),
    html.Br(),
    dbc.Row([
        dbc.Col([], width=5),
        dbc.Col([
            dbc.Button("Show results",
                       id = "worldviewer-submit-button",
                       color = "primary")],
            width=4)])])])
```

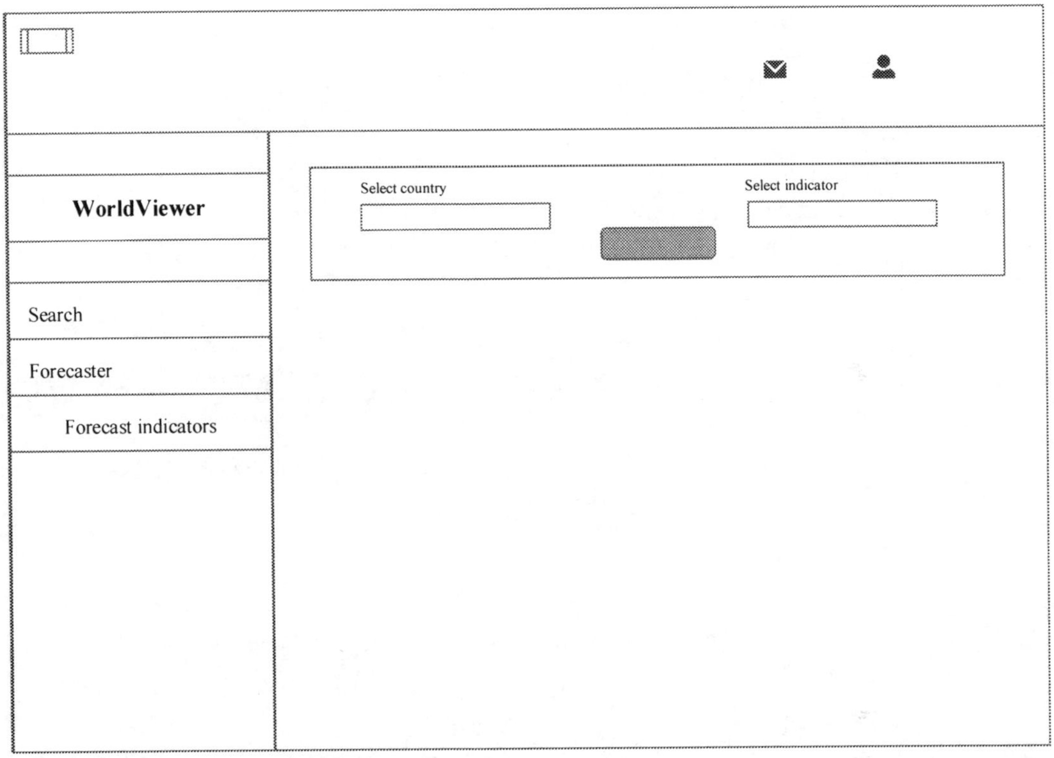

WorldViewer

Search

Forecaster

Forecast indicators

Select country

Select indicator

Figure 8-8. *Search bar*

Creating Interactive Charts

To streamline the data for a user, contain interactive charts in the app.

Listing 8-18 develops a plain chart that updates when a user selects a country and an indicator and clicks the "Show results" button by implementing the `Graph()` method from the `dash_core_components` library and containing it in the `Card()` method from the `dash_bootstrap_components`.

The `callback` segment of this chapter updates the chart. To ensure that the chart is plain prior to a user taking action, it specifies `displayModeBar` as `False`.

Figure 8-9 is an example of the output when a user selects a country and an indicator and clicks the "Shows results" button.

Listing 8-18. Containing an Interactive Chart

```
INTERACTIVE_CHARTS = dbc.Row([
    dbc.Col([
        dbc.Card([
            dbc.CardBody([
                dbc.Row([
                    dbc.Col([
                        dcc.Graph(id = "worldviewer-lineplot",
                                  figure = {"data" : [{"x" : [1 , 2], "y" :
                                  [3 , 1]}]},
                                  config = {"displayModeBar" : False})],
                        width = 12)]),
                dbc.Row([
                    dbc.Col([
                        dcc.Graph(id = "worldviewer-histogram",
                                  figure = {"data" : [{"x" : [1 , 2], "y" :
                                  [3 , 1]}]},
                                  config = {"displayModeBar" : False})],
                        width = 6),
                    dbc.Col([
                        dcc.Graph(id = "worldviewer-boxplot",
                                  figure = {"data" : [{"x" : [1 , 2], "y" :
                                  [3 , 1]}]},
                                  config = {"displayModeBar" : False})],
                        width = 6)],
                    align = "center")])],
            style={"width" : "auto"})],
        width=12)],
    align = "center")
```

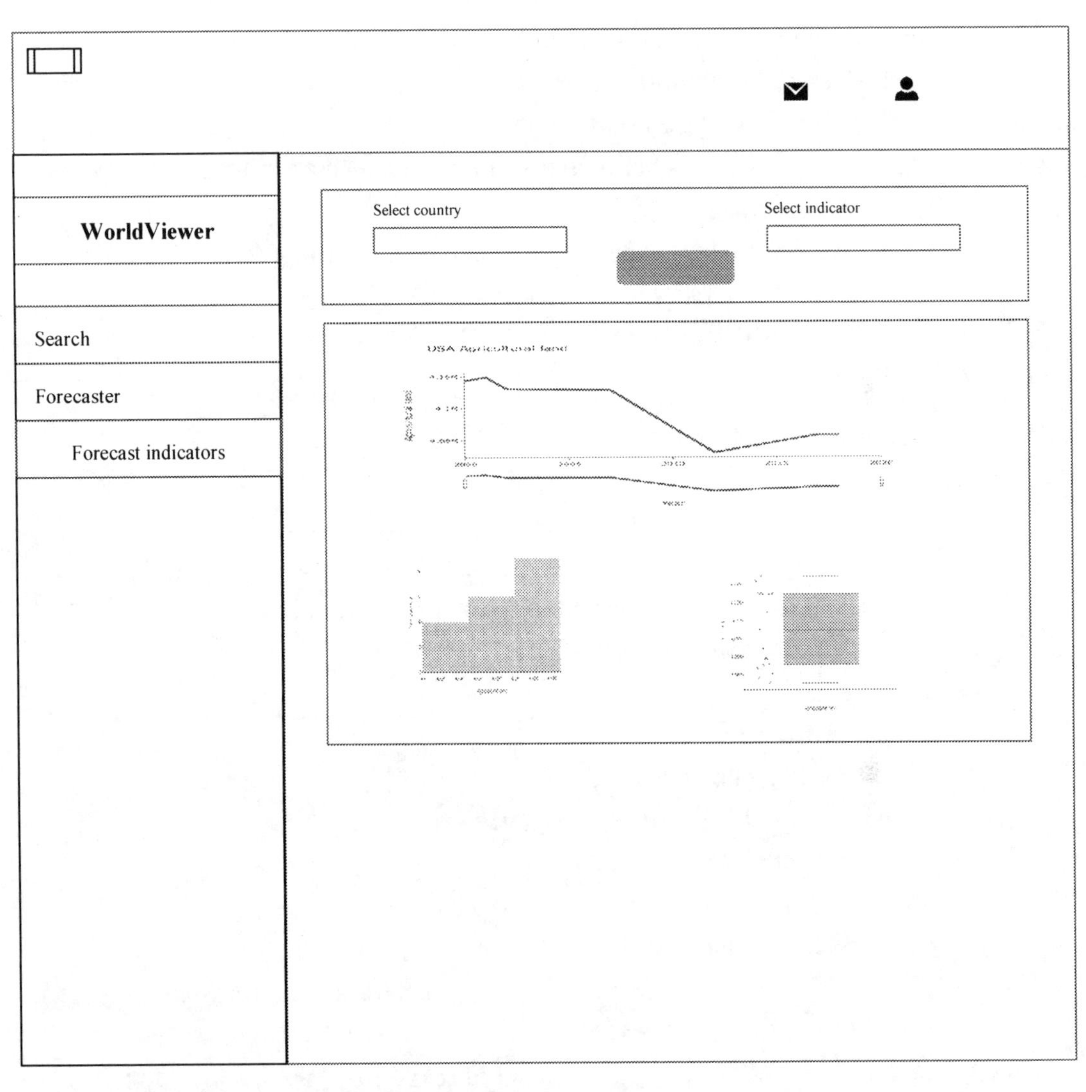

Figure 8-9. *Containing an interactive chart*

Containing an Interactive Table and Allowing Generating a Report and Enabling Download

Another prevalent way of displaying data involves tabulating it. Listing 8-19 develops an empty component by implementing the `Div()` method from the `dash_html_components` library and specifies `id = "worldviewer-table"`. Subsequently, it develops another `Div()` method with `id = "share-market-series-and-data"`, containing the interactive chart in Listing 8-18 and the table. The `callback` segment updates the table when a user selects a country, indicator, and the "Show results" button.

It develops a button that generates a report and enables downloading by implementing the `Button()` method from the `dash_bootstrap_components` library and specifies `id = "download-original-data-button"`. The `Download()` method from the `dash_extensions` library, which is beneath the button (`id = "download-worldviewer-results"`) facilitates the download function.

Figure 8-10 is an example of the output when a user clicks the "Show Data" button.

Listing 8-19. Containing an Interactive Table and Generating a Report and Enabling Download

```
INTERACTIVE_TABLE = dbc.Row([
    dbc.Col([
        dbc.Button(id = "collapsemarket-worldviewer-results-button",
                   n_clicks = 0,
                   children = "Show descriptive statistics",
                   className = "mr-1",color="primary"),
        dbc.Collapse(
            dbc.Card([
                dbc.CardBody([
                    html.Div(id = "worldviewer-table"),
                    html.Br(),
                    dbc.Row([
                        dbc.Col([
                             dbc.Button(id = "download-worldviewer-results-
                             button",
                                       n_clicks = 0,
                                       children = "Download descriptive
                                       statistics",
                                       className = "mr-
                                       1",color="primary"),
                            Download(id = "download-worldviewer-results")],
                            width = 3)])])],
                style={"width" : "12",
                       "paper_bgcolor" : "rgba(0,0,0,0)"}),
            id = "collapse-worldviewer-results")],
        width = 12)])
```

Listing 8-20 merges the components using the Div() method.

Listing 8-20. Merge Components

```
WORDVIEWER_SUMMARY = html.Div([
    dbc.Row([dbc.Col([INTERACTIVE_CHARTS])]),
    dbc.Row([
        dbc.Col([INTERACTIVE_TABLE])])],
    id="collapse-worldviewer-page")
```

Listing 8-21 completes the search navigation bar using the Div() method.

Listing 8-21. Complete the Search Navigation Bar

```
WORDVIEWER_LAYOUT = html.Div([
    html.Br(),
    dbc.Row([
        dbc.Col([INPUT_CARD],width=12)]),
    html.Br(),
    dbc.Row([
        dbc.Col([],width=5),
        dbc.Col([])]),
    dbc.Collapse([WORDVIEWER_SUMMARY],
                id="collapse-worldviewer-menu")])
```

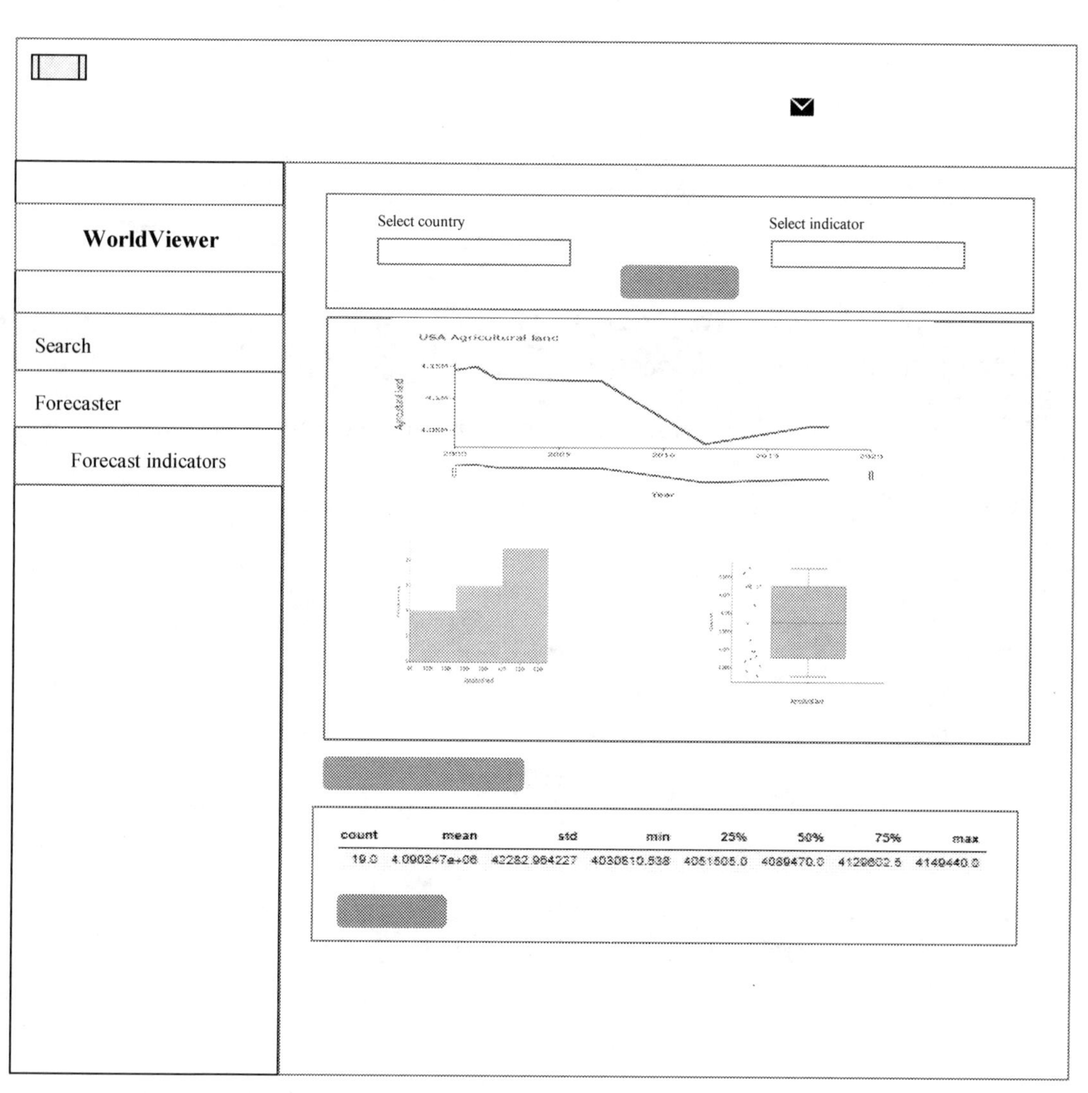

Figure 8-10. *Containing an interactive table*

Specifying the App Layout

After completing the Dash app design, specify the app layout. To do so, use `app.layout` with a `Div()` method that contains the `Store()` and `Location()` method to make the side navigation bar responsive.

Listing 8-22 specifies the Dash app layout.

Listing 8-22. Specifying the App Layout

```
content = html.Div(
    id = "app-content",
    style=APP_CONTENT_STYLE)
app.layout = html.Div(
    [
        dcc.Store(id = "responsive-sidebar-click"),
        dcc.Location(id = "url"),
        navbar,
        RESPONSIVE_SIDE_NAVIGATION_BAR,
        content,
    ],
)
```

Specifying a Callback Function

The most crucial part of a Dash web app is a *callback*, which works similarly to the "`get-post`" method. It comprises three key dependencies, namely, `Input()` method for specifying input, `State()` method for specifying the state undertaking an action (through which the `Input()` method depends on for action), and `Output()` method for specifying the output.

Figure 8-11 simplifies callback functions, which constructs a side navigation bar that hides and unhides when a user clicks the toggle button.

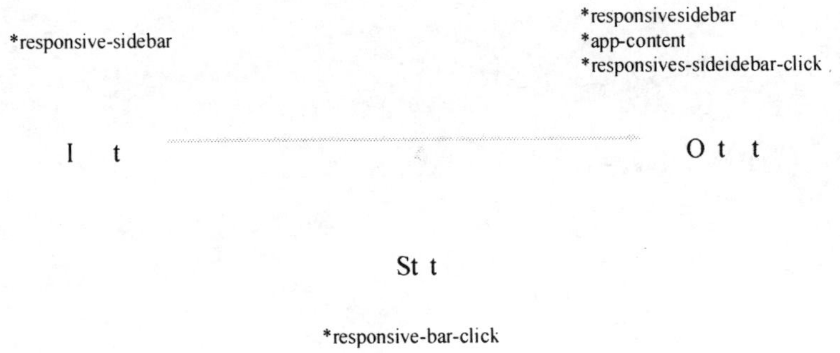

*responsive-sidebar

*responsivesidebar
*app-content
*responsives-sideidebar-click

I t

O t t

St t

*responsive-bar-click

Figure 8-11. *Callback function example*

Figure 8-11 shows that when a user clicks the toggle button, the side navigation bar hides or unhides.

A callback code begins with `@app.callback()` and contains the `Input()`, `State()`, and `Output()` methods in it. For each method, specify the component ID and argument for action.

Listing 8-23 shows the component (toggle button) ID in the `Input()` method is "`btn_sidebar`" and the argument is "`n_clicks`" (which represents the number of clicks). The IDs for the `Output()` methods are "`sidebar`", "`page-content`", and "`side_click`". The arguments for the first two are "`style`" and the last one is "`data`".

Tip Specify callbacks after specifying app.layout. Conclude a callback by specifying the define function as `def (): return`.

Callback for a Responsive Side Navigation Bar

Listing 8-23 constructs a `callback` for hiding and unhiding the side navigation bar based on a user clicking the toggle button. Note that the `def()` function comprises conditional statements (`if else`) to hide and unhide the side navigation bar.

Listing 8-23. Callback for a Responsive Side Navigation Bar

```
@app.callback(
    [
        Output("responsivesidebar", "style"),
        Output("app-content", "style"),
        Output("responsive-sidebar-click", "data"),
    ],

    [Input("toggle-button", "n_clicks")],
    [
        State("responsive-sidebar-click", "data"),
    ]
)
```

```
def toggle_responsivesidebar(n, nclick):
    if n:
        if nclick == "SHOW":
            RESPONSIVE_MENU_style = RESPONSIVE_MENU_HIDEN
            APP_CONTENT_style = APP_CONTENT_STYLE1
            NO_OF_CURRENT_CLICKS = "HIDDEN"
        else:
            RESPONSIVE_MENU_style = RESPONSIVE_MENU_STYLE
            APP_CONTENT_style = APP_CONTENT_STYLE
            NO_OF_CURRENT_CLICKS = "SHOW"
    else:
        RESPONSIVE_MENU_style = RESPONSIVE_MENU_STYLE
        APP_CONTENT_style = APP_CONTENT_STYLE
        NO_OF_CURRENT_CLICKS = "SHOW"
    return RESPONSIVE_MENU_style, APP_CONTENT_style, NO_OF_CURRENT_CLICKS
def toggle_collapse(n, is_open):
    if n:
        return not is_open
    return is_open
def set_navitem_class(is_open):
    if is_open:
        return "open"
    return ""
```

Callback for URL Routing

It is unwise to have all the source code in one `.py` file when working on large-scale web app developments. It makes it difficult to manage the project. In addition, a slight error in one line of code affects the entire file. The best way to address this issue is to have different pages in different `.py` files and construct a connection between them through URL routing, thus specifying a URL link that takes a user to a specific area within the app.

Listing 8-24 specifies URL routing by implementing the `callback()` method that contains an `Input()` method with the ID as "`url`" and argument as "`pathname`", and an `Output()` method with the ID as "`page-content`" and argument as "`children`". Likewise, it comprises a `def()` function containing conditional statements (`if elif`), which directs the user to a specific URL. If a user specifies an invalid URL, a 404 message appears.

Listing 8-24. Callback for URL Routing

```
path_name_map = {"/": WORDVIEWER_LAYOUT,
                 "/page-1/1": WORDVIEWER_LAYOUT,
                 "/page-2/1": "Forecast Indicators",
                 "/page-3/1": "Profile",
                 "/page-4/1": "Edit Profile",
                 "/page-4/2": "Privacy & Safety",
                 "/page-4/3": "Account Settings",
                 "/page-4/4": "Billing",
                 "/page-4/5": "Sign Out"}
for i in range(0, 4):
    app.callback(
        Output(f"responsivemenu-{i}-collapse", "is_open"),
        [Input(f"responsivemenu-{i}", "n_clicks")],
        [State(f"responsivemenu-{i}-collapse", "is_open")],
    )(toggle_collapse)
    app.callback(
        Output(f"responsivemenu-{i}", "className"),
        [Input(f"responsivemenu-{i}-collapse", "is_open")],
    )(set_navitem_class)
@app.callback(Output("app-content", "children"), [Input("url",
"pathname")])
def render_page_content(pathname):
    return html.P(path_name_map[pathname])
```

Specifying a Callback Function for Unhiding Content

Listing 8-25 specifies a callback function for unhiding content upon the user clicking the "Show results" button.

Listing 8-25. Specifying a Callback Function for an Unhiding Content

```
@app.callback(Output("collapse-worldviewer-menu", "is_open"),
              [Input("worldviewer-submit-button", "n_clicks")])
def toggle_collapse_worldviewer_menu(n):
    return n
```

Specifying a Callback Function for Interactive Charts

This section develops a callback to update charts when a user selects a country and an indicator and clicks the search button. There are three charts; therefore, there are three callback functions.

Note that the def() function extracts the stock data from the pandas-datareader library with a predefined date as today's date.

Listing 8-26 specifies a callback function for an interactive line plot.

Listing 8-26. Specifying a Callback Function for an Interactive Line Plot

```
@app.callback(Output("worldviewer-lineplot", "figure"),
              [Input("country-indicator", "value")],
              [State("country-symbol", "value")])
def draw_worldviewer_lineplot(indicator, country):
    df = wb.download(indicator = indicator, country = [country], start =
    2000, end = 2021)
    df_country = pd.DataFrame(df)
    df_country = df_country.reset_index()
    country_name = [df_country.country[0]]
    ind = wbdata.get_indicator(indicator)
    text = ind[0]["name"]
    title = text[:text.find("(")-1]
    df.columns = [title]
    figure = go.Figure(data=go.Scatter(x = df_country.year,
                                       y = df_country.iloc[::,-1],
                                       mode = "lines",
                                       line = dict(color = "#1266F1",
                                                   width = 4)))
    figure['layout'] = {"title":  "".join(country_name) + " " +
    "".join(title),
                        "xaxis": {"anchor": "y", "domain": [0.0, 1.0],
                                  "title": "Year"},
                        "yaxis": {"anchor": "x", "domain" : [0.0, 1.0],
                                  "title": "".join(title)}}
```

```
    figure['layout'].update(autosize=True,
                            template="simple_white",
                            showlegend=False)
    figure.update_xaxes(rangeslider_visible=True)
    return figure
```

Listing 8-27 specifies a callback function for an interactive histogram.

Listing 8-27. Specifying a Callback Function for an Interactive Histogram

```
@app.callback(Output("worldviewer-histogram", "figure"),
              [Input("country-indicator", "value")],
              [State("country-symbol", "value")])
def draw_worldviewer_histogram(indicator, country):
    df = wb.download(indicator = indicator, country = [country], start =
    2000, end=2021)
    df_country = pd.DataFrame(df)
    df_country = df_country.reset_index()
    country_name = [df_country.country[0]]
    ind = wbdata.get_indicator(indicator)
    text = ind[0]["name"]
    title = text[:text.find("(")-1]
    df.columns = [title]
    figure = px.histogram(df,
                          color_discrete_sequence = ["lightskyblue"])
    figure['layout'] = {"title":  "",
                        "xaxis": {"anchor": "y", "domain": [0.0, 1.0],
                                  "title": "".join(title)},
                        "yaxis": {"anchor": "x", "domain" : [0.0, 1.0],
                                  "title": "Frequency"}}
    figure["layout"].update(showlegend=False)
    return figure
```

Listing 8-28 specifies a callback function for an interactive box plot.

Listing 8-28. Specifying a Callback Function for an Interactive Box Plot

```
@app.callback(Output("worldviewer-boxplot", "figure"),
              [Input("country-indicator", "value")],
              [State("country-symbol", "value")])
def draw_worldviewer_boxplot(indicator, country):
    df = wb.download(indicator = indicator, country = [country], start =
    2000, end = 2021)
    df_country = pd.DataFrame(df)
    df_country = df_country.reset_index()
    country_name = [df_country.country[0]]
    ind = wbdata.get_indicator(indicator)
    text = ind[0]["name"]
    title = text[:text.find("(")-1]
    df.columns = [title]
    figure = px.box(df,
                    color_discrete_sequence=["darkorange"],
                    points="all")
    figure["layout"].update(showlegend=False)
    figure['layout'] = {"title":  "",
                        "xaxis": {"anchor": "y", "domain": [0.0, 1.0],
                                  "title": ""},
                        "yaxis": {"anchor": "x", "domain" : [0.0, 1.0],
                                  "title": "Count"}}
    return figure
```

Specifying a Callback Function for Unhiding an Interactive Table

Listing 8-29 specifies a callback function for unhiding an interactive table.

Listing 8-29. Specifying a Callback Function for Unhiding an Interactive Table

```
@app.callback(Output("collapse-worldviewer-results", "is_open"),
              [Input("collapsemarket-worldviewer-results-button",
              "n_clicks")],
              [State("collapse-worldviewer-results", "is_open")])
```

```
def toggle_collapse_worldviewer_results_table(n, is_open):
    if n:
        return not is_open
    return is_open
```

Specifying a Callback Function for an Interactive Table

Listing 8-30 develops a callback to update the table when a user selects a country and an indicator and clicks the "Show descriptive statistics" button. Note that the def() function extracts the indicator data from the pandas-datareader library with a predefined date as today's date.

To update the table, it uses the dash_table library.

Listing 8-30. Specifying a Callback Function for an Interactive Table

```
@app.callback(Output("worldviewer-table", "children"),
              [Input("country-indicator","value")],
              [State("country-symbol","value")])
def draw_worldviewer_table(indicator, country):
    df = wb.download(indicator = indicator, country = [country],
    start = 2000, end = 2021)
    df_country = pd.DataFrame(df)
    df_country = df_country.reset_index()
    country_name = [df_country.country[0]]
    ind = wbdata.get_indicator(indicator)
    text = ind[0]["name"]
    title = text[:text.find("(")-1]
    df.columns = [title]
    descriptive_statistics = df.describe().transpose()
    data = descriptive_statistics.to_dict("rows")
    columns = [{"name": i, "id": i,} for i in (descriptive_statistics.
    columns)]
    return dt.DataTable(data=data,columns=columns,style_table={"overflow":
    "auto",
                                                    "striped":"True",
                                                    "bordered":"True",
                                                      "hover":"True"})
```

Specifying a Callback Function for Callback for Data Download

Listing 8-31 develops a callback to download a Microsoft Excel file when a user clicks the "Download descriptive statistics" button. Note that the `def()` function extracts the indicator data from the `pandas-datareader` library with a predefined date as today's date. It also applies the `send_frame()` method from the `dash_extensions` library to enable downloads.

Listing 8-31. Specifying a Callback Function for Callback for Data Download

```
@app.callback(Output("download-worldviewer-results", "data"),
              [Input("download-worldviewer-results-button", "n_clicks")],
              [State("country-indicator","value"),
               State("country-symbol","value")])
def download_worldviewer_results_data(n_clicks, indicator, country):
    df = wb.download(indicator = indicator, country = [country], start =
    2000, end = 2021)
    df_country = pd.DataFrame(df)
    df_country = df_country.reset_index()
    country_name = [df_country.country[0]]
    ind = wbdata.get_indicator(indicator)
    text = ind[0]["name"]
    title = text[:text.find("(")-1]
    df.columns = [title]
    describe = df.describe().transpose()
    return send_data_frame(df.to_excel,
                           filename = "".join(country_name) + " " +
                           "".join(title) + " descriptive statistics.xlsx")
```

Run the Dash App

Listing 8-32 runs the Dash app by implementing the `run_server()` method and specifying mode as "`external`", including `dev_tools_ui` and `dev_tools_props_check` as `False` so that it does not debug the app prior to running it.

Listing 8-32. Specifying a Callback Function for URL Routing

```
app.run_server(mode="external",
               dev_tools_ui = False,
               dev_tools_props_check = False)
```

Conclusion

This chapter introduced a functional approach to creating a web application comprising a top and side navigation bar that responds to user input. First, it introduced a technique for attaining a CSS script to make use of icons. Then, it presented an approach to creating icons with a hyperlink, thus enabling a user to move from one page to another. Afterward, it showed tactical submenus in the sidebar. Besides that, it revealed a way to collapse items upon clicking a specific component. Most importantly, it familiarizes you with the `callback()` method to facilitate reactions.

CHAPTER 9

Basic Web App Authentication

If you intend on sharing a Dash web app across users, especially in an organizational context, ensure that you amply secure the app and restrict certain functionalities based on the user level. The considerable complexity of user management and privacy varies tremendously from one organization to another, one project to another; thus, there is no universal approach to user authentication.

This chapter is not prescriptive; rather, it valiantly attempts to acquaint you with basic web app user authentication. It exhibits the fundamentals by implementing key Python web frameworks (i.e., `dash_auth` and `flask`). Besides that, it presents schemes for building authentication inputs by implementing `dash_core_components`. It reasonably concludes by pointing out significant resources relating to Dash web app user authentication.

Authentication with Dash Auth

The Dash library's `dash_auth` module makes web app authentication relatively easy. Ensure that you have `dash_auth` installed on your environment. To install it in a Python environment, use `pip install dash-auth`. To install it on conda, use `conda install -c conda-forge dash-auth`.

Listing 9-1 structures a basic Dash web app authentication (see Figure 9-1). Note that storing the username and password inside the `.py` file is not the most secure way to approach app security; it was done here for demonstration purposes.

T. C. Nokeri, *Web App Development and Real-Time Web Analytics with Python*,
https://doi.org/10.1007/978-1-4842-7783-6_9

Listing 9-1. Dash Basic Authentication

```
import dash
import dash_auth
import dash_html_components as html
user_details = [
    ["Tshepo", "Tshepo!Password!#$897"]
]
external_stylesheets = ['https://codepen.io/chriddyp/pen/bWLwgP.css']
app = dash.Dash(__name__, external_stylesheets=external_stylesheets)
auth = dash_auth.BasicAuth(
    app,
    user_details
)
app.layout = html.Div(
    [
        html.P("Login successful")
    ],
    className = "container")
app.scripts.config.serve_locally = True
if __name__ == '__main__':
    app.run_server(debug=False)
```

Figure 9-1. *Basic authentication*

Figure 9-1 shows a pop-up that requires a user to enter the username and password before using the web app.

Alternatively, upgrade to Dash Enterprise to easily configure security features with no code. Learn more about Dash Enterprise at the official Plotly website (`https://dash.plotly.com/dash-enterprise`).

Authentication with Flask

This next segment demonstrates an approach for basic Dash web app authentication by implementing the Flask library. First, ensure that you have installed it in your environment. To install the Flask library in a Python environment, use `pip install flask`. To install it in a conda environment, use `conda install -c conda-forge flask`.

Next, create a `.py` file that holds username and password, then build another one containing the functionality (the form with inputs and a `get-post` method for app routing).

```
user_profile.py
authentication.py
```

Listing 9-2 presents code in the `user_profile.py` file that contains both the username and password. It implements Fernet from the cryptography library to generate a secret key using the `generate_key()` method. Following that, it opens a `.bin` file containing the password. Subsequently, it encrypts and decodes the password. Learn more about cryptography at `https://cryptography.io/en/latest/fernet/`.

Listing 9-2. Specifying Usernames and Passwords

```
import pandas as pd
import io
key = Fernet.generate_key()
fernet_key = Fernet(key)
password_file = r"filepath\password.bin"
with open(password_file, mode='rb') as file:
    password = file.read()
encrypted_token = fernet_key.encrypt(password)
decrypted_token = fernet_key.decrypt(encrypted_token)
user_password = decrypted_token.decode()
user_name_file = r"filepath\user__name.bin"
with open(user_name_file, mode='rb') as file:
    user_names = file.read()
def users_details():
    return user_password, user_names
```

Other libraries that support password encryption, including werkzeug (https://werkzeug.palletsprojects.com/en/2.0.x/utils/#module-werkzeug.security). Note that a .bin file was used for demonstration purposes; preferably, use a secured database.

Listing 9-3 imports Flask and the function specified as "users_details" in the user_profile.py file. Following that, it specifies the app route and conditional statements for authentication.

Listing 9-3. Authentication with Flask

```
import flask
import dash_html_components as html
import dash_core_components as dcc
from Jupyter_dash import JupyterDash
from user_profile import users_details
user_password, user_names = user_profile()
app = JupyterDash(external_stylesheets=[dbc.themes.BOOTSTRAP])
CONTENT_STYLE = {"margin-left": "16rem",
                 "padding": "0.5rem 0.5rem",
                 "color": "gray"}
sign_in_form = html.Div([
    html.Form([
        dcc.Input(placeholder="Enter username",
                  name="username",
                  type="email"),
        dcc.Input(placeholder="Enter password",
                  name="password",
                  type="password"),
        html.Button("Login", type='submit')],
        action="/login",
        method="post")])
_app_route = ""/""
@app.server.route("/login", methods=["POST"])
def routes():
    data = flask.request.form
    username = data.get("username")
    password = data.get("password")
```

```
    if username not in user_password.keys() or  user_password[username] !=
    password:
        return flask.redirect("/login")
    else:
        return_sess = flask.redirect(_app_route)
        return_sess.set_cookie("custom-auth-session", username)
        return return_sess
content = html.Div(id="page-content",
                   style=CONTENT_STYLE)
app.layout = html.Div([sign_in_form,
                       content])
app.run_server(mode='external',
               dev_tools_ui=False,
               dev_tools_props_check=False)
```

Login Form

Alternatively, construct a separate page for logging on to a web app. Listing 9-4 creates a form on a login page (see Figure 9-2).

Listing 9-4. Login Form

```
app = JupyterDash(external_stylesheets = [dbc.themes.MATERIA],
                  meta_tags = [{"charset":"utf-8",
                                "name": "viewport",
                                "content": "width=device-width, initial-
                                scale=1"}])
CONTENT_STYLE = {"margin-left": "16rem",
                 "padding": "0.5rem 0.5rem",
                 "color": "gray"}
user_email_input = dbc.FormGroup([
    dbc.Label("Email",
              html_for = "user-email-input",
              width = 4),
```

```
    dbc.Col([
        dbc.Input(type = "email",
                  id = "user-email-input",
                  placeholder = "Enter email address")],
        width = 8)],
    row = True)
user_password_input = dbc.FormGroup([
    dbc.Label("Password",
              html_for ="user-password-input",
              width = 4),
        dbc.Col([
            dbc.Input(type = "password",
                      id = "user-password-input",
                      placeholder = "Enter password")],
            width = 8)],
    row = True)
sign_in_button = html.Div([
    dbc.Button("Login",
               color = "primary",
               id = "user-login-input")])
remember_password_input = dbc.FormGroup([
    dbc.Label("",
              html_for = "remember-password-checklist",
              width = 2),
        dbc.Col([
            dbc.Checklist(id = "remember-password-checklist",
                          options = [{"label": "Remember Password",
                                      "value": 1}],
                          switch = True )
        ],
            width = 10)],
    row = True)
forgot_password = html.Label(["",
                              html.A("Create an account",
                                     href = "#")])
```

```
form = dbc.Form([
    dbc.Row([
        dbc.Col( [
            user_email_input
        ]
        )
    ]
    ),
    dbc.Row([
        dbc.Col(
            [
                user_password_input
            ]
        )
    ]
    ),
    dbc.Row([
        dbc.Col(
            [
                remember_password_input
            ]
        )
    ]
    ),
    dbc.Row([
        dbc.Col([], width = 5),
            dbc.Col(
                [
                    sign_in_button
                ]
            )
    ]
    )
]
)
```

```
sign_in_form = dbc.Card([
    dbc.CardBody([
        dbc.Row([
            dbc.Col(
                [
                    html.H4("Sign in",
                            style = {"text-align":"center"}),
                    form
                ]
            )
        ]
        ),
        html.Br(),
        dbc.Row([
            dbc.Col(
                [
                    forgot_password
                ]
            )
        ]
        )
    ]
    )
 ],
     style = {"margin-left": "28rem",
              "margin-right": "28rem",
              "margin-top":"8rem",
              "margin-bottom":"8rem",
              "padding": "0.5rem 0.5rem",
              "background-color": "#FFFFFF"},
     className = "card-header")
 copyrights = dbc.Card([
     dbc.CardBody([
         html.Br(),
         dbc.Row([
```

```
            dbc.Col([], width = 5),
            dbc.Col(
                [
                    html.P("Copyright © 2021 WorldViewer. All rights
                    reserved.",
                           style = {"color":"dark"}
                          )
                ]
            )
        ]
        )
    ]
    )
],
    color="light")
sign_in_jumbotron = dbc.Jumbotron(
    [
        sign_in_form
    ]
)
content = html.Div(id="page-content",
                   style = CONTENT_STYLE)
app.layout = html.Div(
    [
        sign_in_jumbotron,
        copyrights,
        content
    ]
)
app.run_server(mode = "external",
               dev_tools_ui = False,
               dev_tools_props_check = False)
```

Username

Password

Figure 9-2. *Login form*

Figure 9-2 shows an input type for entering a username and another for entering the password, including a login button. The components may be configured to trigger the authentication of a user's profile using some user database.

Login on Home Page

At times, you may want to place the login form on the homepage to make it easier for users to log in to a dashboard or some part of a web app. Listing 9-5 creates a login section on the home page (see Figure 9-3).

Listing 9-5. Login on Home Page

```
app = JupyterDash(external_stylesheets=[dbc.themes.BOOTSTRAP])
user_email_input = dbc.FormGroup([
    dbc.Label("Email",
              html_for = "user-email-input",
              width = 3),
    dbc.Col([
        dbc.Input(type ="email",
                  id = "user-email-input",
                  placeholder = "Enter email")
    ],
```

```
        width = 10
    )
],
    row = True)
user_password_input = dbc.FormGroup([
    dbc.Label("Email",
              html_for = "user-password-input",
              width = 3),
    dbc.Col(
        [
            dbc.Input(type="password",
                      id="user-password-input",
                      placeholder="Enter email"
                     )
        ],
        width = 10
    )
],
    row = True)
sign_in_button = html.Div(
    [
        dbc.Button("Login",
                   color="primary",
                   id="user-login-input"
                  )
    ]
)
login_div = html.Div(
    [
        dbc.Row(
            [
                dbc.Col(
                    [
                        user_email_input,
                    ]
                ),
```

```
                    dbc.Col(
                        [
                            user_password_input,
                            sign_in_button]
                    )
                ]
            )
        ]
)
navigation_bar_item_1 = dbc.NavItem(
    [
        dbc.NavLink("Home",
                    href = "#")
    ]
)
navigation_bar_item_2 = dbc.NavItem(
    [
        dbc.NavLink("What We Do",
                    href = "#")
    ]
)
navigation_bar_item_3 = dbc.NavItem(
    [
        dbc.NavLink("Solutions",
                    href = "#")
    ]
)
navigation_bar = dbc.NavbarSimple(
    children=[navigation_bar_item_1,
              navigation_bar_item_2,
              navigation_bar_item_3,
              login_div],
    brand="Worldviewer",
    brand_href="#",
    sticky="top",
```

```
    className="mb-5",
    color="light")
content = html.Div(id = "page-content")
app.layout = html.Div(
    [
        navigation_bar,
        content
    ]
)
app.run_server(mode="external",
               dev_tools_ui=False,
               dev_tools_props_check=False)
```

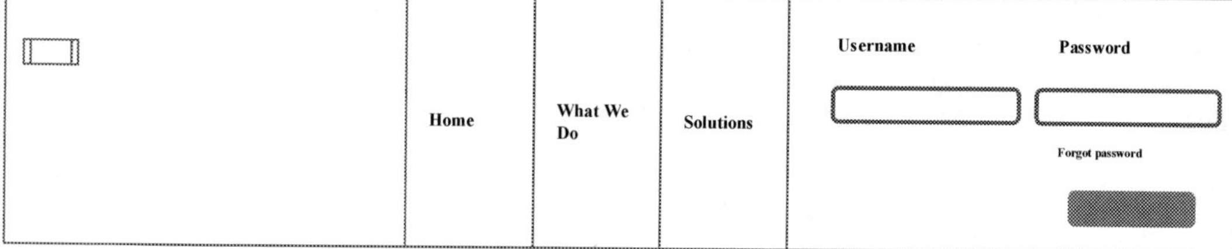

***Figure 9-3.** Log in on home page*

Conclusion

This chapter introduced the essentials of basic authentication using the Dash library. For more security features, you may incorporate other web APIs, such as Auth0 (`https://auth0.com/authenticate/python/amazon/`) or Okta (`https://developer.okta.com/code/python/`), and security features that come with cloud services, like Microsoft Azure, Amazon Web Services, or IBM Cloud.

I suggest you look at Dash Enterprise (`https://plotly.com/dash/authentication/`), which makes it relatively easy to set up security features in a Dash web app without explicit programming.

CHAPTER 10

Dash into a Full Website

Prior chapters introduced building dashboards as web applications integrated with machine learning models. This chapter takes it a step further. After reading the contents of this chapter, you should be able to build important pages of a web app.

Listing 10-1 installs libraries that the chapter employs (i.e., `dash`, `dash_core_components`, `dash_html_components`, `dash_bootstrap_components`, and `jupyter_dash`), including Dash dependencies (`Input`, `Output`, and `State`).

Listing 10-1. Import Key Dependencies

```
import dash_table as dt
import dash
import dash_core_components as dcc
import dash_html_components as html
from dash.dependencies import Input, Output, State
import dash_bootstrap_components as dbc
from jupyter_dash import JupyterDash
```

Home Page

A home page is often the main page of a website. It provides customers/users with basic information relating to an organization/individual. For organizations, it comprises a navigation bar that contains links to the About Us page, products/services page. Sometimes it may contain a login section. Then, there is a header containing any important messages, products, services, and sections containing other information and media.

This section explains how to build a home page step by step. You individually create each component so that you can reuse it.

T. C. Nokeri, *Web App Development and Real-Time Web Analytics with Python*,
https://doi.org/10.1007/978-1-4842-7783-6_10

Listing 10-2 initializes the dash app by implementing the `JupyterDash()` method from the JupyterDash library. This library is useful in prototyping with a Jupyter Notebook. However, for production development, I suggest you use the original Dash library (see Listing 10-3).

Listing 10-2. Initialize Dash App Using JupyterDash

```
font_awesome = "https://use.fontawesome.com/releases/v5.8.1/css/all.css"
app = JupyterDash(external_stylesheets=[dbc.themes.MATERIA, font_awesome],
                  meta_tags=[{"charset":"utf-8",
                              "name": "viewport",
                              "content": "width=device-width,
                              initial-scale=1"}])
```

Listing 10-3. Initialize Dash App Using the Original Dash Library

```
app = dash.Dash(external_stylesheets=[dbc.themes.MATERIA, font_awesome],
                meta_tags=[{"charset":"utf-8",
                            "name": "viewport",
                            "content": "width=device-width, initial-
                            scale=1"}])
```

Listing 10-4 specifies the style of the page using the CSS format.

Listing 10-4. Specify Style Using CSS

-

Listing 10-5 constructs navigation items by implementing the `NavItem()` method. Then it specifies the name and link (using `href`) the item contains by implementing the `NavLink()` method. Navigation items include "`Home`", "`What We Do`", and "`Solutions`".

Listing 10-5. Navigation Bar Items

```
nav_item1 = dbc.NavItem(dbc.NavLink("Home",
                                    href = "#"))
nav_item2 = dbc.NavItem(dbc.NavLink("What We Do",
                                    href = "#"))
nav_item3 = dbc.NavItem(dbc.NavLink("Solutions",
                                    href = "#"))
```

In addition to the navigation items specified in Listing 10-5, let's include other navigation items, including one to release the DropdownMenu() method (see Figure 10-1).

Listing 10-6. Navigation Bar Items with a Drop-Down Menu

```
dropdown_solutions = dbc.DropdownMenu(
    children=[
        dbc.DropdownMenuItem("Economic Analysis"),
        dbc.DropdownMenuItem(divider = True),
        dbc.DropdownMenuItem("Environment Analysis"),
        dbc.DropdownMenuItem(divider = True),
        dbc.DropdownMenuItem("Forecasting"),
    ],
    nav = True,
    in_navbar = True,
    label="Solutions",
    color="dark",
)
```

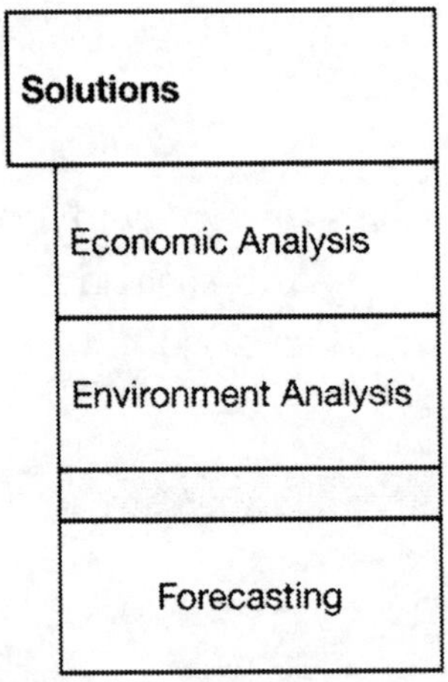

Solutions
Economic Analysis
Environment Analysis
Forecasting

Figure 10-1. *Navigation item with a drop-down menu*

Listing 10-7 constructs cards by implementing the Card() method. Each card contains some insight into what the overall organization is about.

Listing 10-7. Creating Cards

```
app_features1 = dbc.Card([
    dbc.CardBody([
        html.H4("Insights", className  = "card-title"),
        html.P("Gain insights into the market and global economy",
                className = "card-text")
    ])],
    className = "col-sm d-flex",)

app_features2 = dbc.Card([
    dbc.CardBody([
        html.H4("Discover", className = "card-title"),
        html.P("Identify, measure economic and social events and factors",
                className = "card-text")
    ])],
    className = "col-sm d-flex")

app_features3 = dbc.Card([
    dbc.CardBody([
        html.H4("Forecast", className = "card-title"),
        html.P("Project future market conditions and trends.",
                className = "card-text")])],
    className = "col-sm d-flex")
```

Listing 10-8 groups cards created in Listing 10-7. The outcome is shown in Figure 10-2.

Listing 10-8. Group Cards

```
app_features = dbc.Card([
    dbc.CardBody([
        dbc.Row([
            dbc.Col([], width = 2),
            dbc.Col([
```

```
                html.H1("App Features"),
                html.Br(className="my-2"),
                html.P("WorldViewers uses cutting-edge to help clients
                understand world.")],
                width = 3)
        ]
        ),
        dbc.Row([
            dbc.Col(
                [
                    app_features1
                ]
            ),
            dbc.Col(
                [
                    app_features2
                ]
            ),
            dbc.Col(
                [
                    app_features1
                ]
            )
        ]
        )
    ]
    )
]
)
```

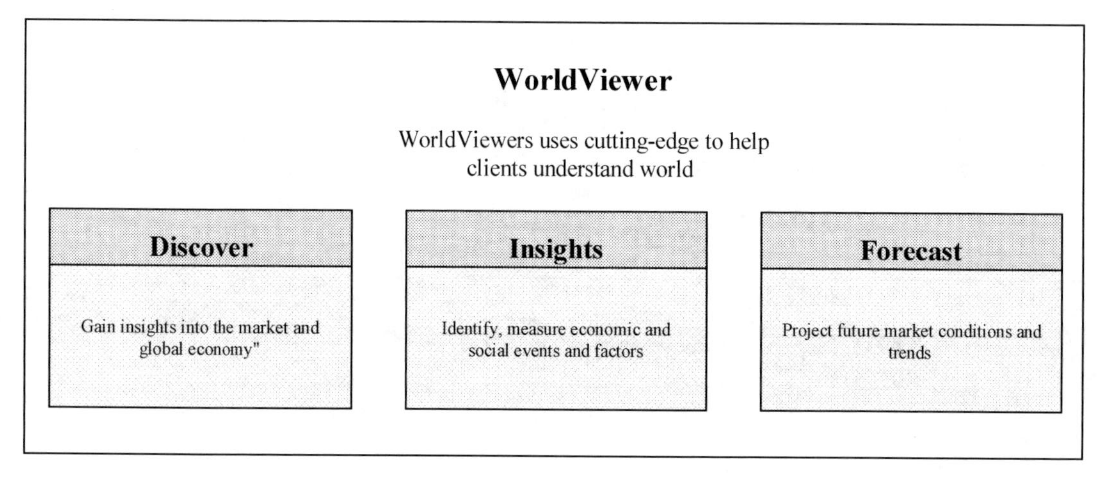

Figure 10-2. *Grouped cards*

Listing 10-9 constructs cards to specify the organization's value proposition (see Figure 10-3). Notice how it applies `Rows()` and `Cols()` to position items.

Listing 10-9. Creating Cards.

```
value_proposition = dbc.Card([
    dbc.CardBody([
        dbc.Row([
            dbc.Col([],width=6),
            dbc.Col([
                html.H2("Economic and Social Analysis on the go.",
                        style = {"color":"white"}),
                html.P("Gain insights and control assets 24/7 on our app,
                so you can experience professional-level economic and
                social analysis features on the move.",
                       style = {"color":"white"})],
                width = 5),
            dbc.Col([],
                    width = "auto")
        ]
        )
    ]
    )
```

```
],
    color="dark"
)
```

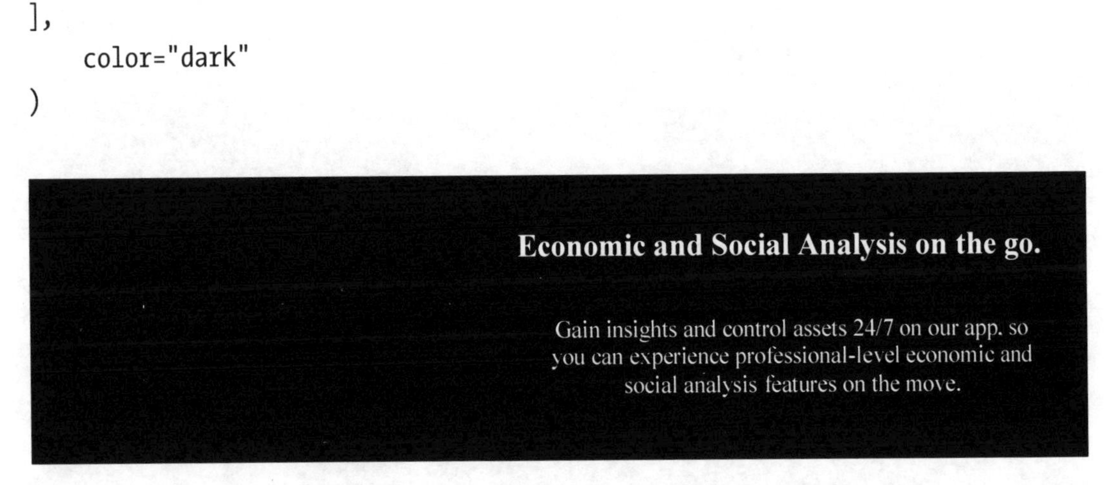

Figure 10-3. *Grouped cards*

Also, Listing 10-10 constructs a card to include a section relating to the organization's offerings.

Listing 10-10. Create Newsletter Subscription Section

```
subscribe_email_input = dbc.FormGroup(
    [
        dbc.Col(
            dbc.Input(
                type = "email",
                id = "subscribe-email",
                placeholder = "Enter email"
            ),
            width = 10,
        ),
    ],
    row = True,
)
subscribe_button = html.Div(
    [
        dbc.Button("Subscribe",
```

```
                    color = "dark",
                    outline = True),
    ]

)
```

Listing 10-11 groups the input and button created in Listing 10-10 (see Figure 10-4).

Listing 10-11. Finalize Subscription Card

```
subscribe_card = dbc.Card([
    dbc.CardBody([
        dbc.Row([
            dbc.Col([], width = 5),
            dbc.Col([
                html.H4("Subscribe to Our Newsletter")],
                width = 5)]),
        dbc.Row([
            dbc.Col([], width = 4),
            dbc.Col(
                [
                    subscribe_email_input
                ],
                width = 4),
        dbc.Col(
            [
                subscribe_button
            ],
            width = 2),
            html.Br()],
            align = "center"),
    ]
    )
],
    style = {"padding": "2rem 1rem"})
```

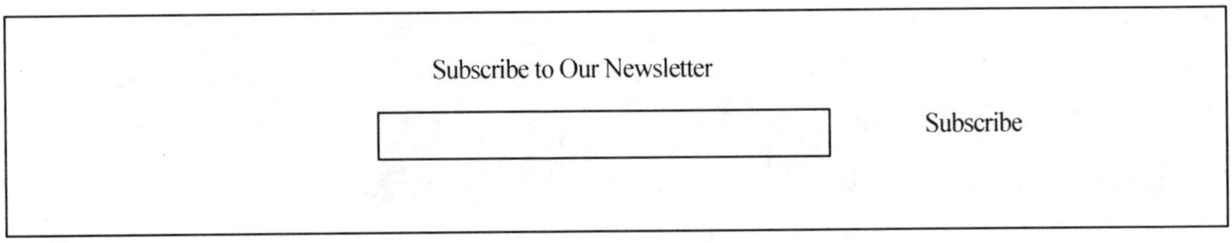

Figure 10-4. *Newsletter subscription*

Listing 10-12 constructs a button for requesting a demo and another for logging on the dashboard (see Figure 10-5).

Listing 10-12. Buttons to Include in the Navigation Menu

```
request_button = html.Div(
    [
        dbc.Button("Request demo",
                   color = "light",
                   outline = False,),
    ]
)
login_button = html.Div(
    [
        dbc.Button("Sign in",
                   color = "primary"),
    ]
)
```

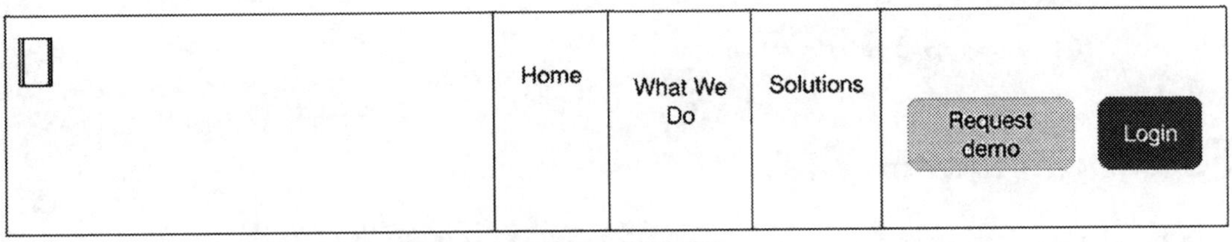

Figure 10-5. *Buttons on navigation bar*

Listing 10-13 constructs navigation constructs a navigation bar that contains navigation items that were specified in the preceding section (i.e., `nav_item1, nav_item2, dropdown_solutions, dropdown_about, dropdown_resources, request_button`, and `login_button`). Figure 10-6 shows the outcome.

Listing 10-13. Navigation Bar

```
default = dbc.NavbarSimple(
    children=[nav_item1,
              nav_item2,
              dropdown_solutions,
              request_button,
              login_button],
    brand="Worldviewer",
    brand_href="#",
    sticky="top",
    className="mb-5",
    color="dark",
    dark=True
)
```

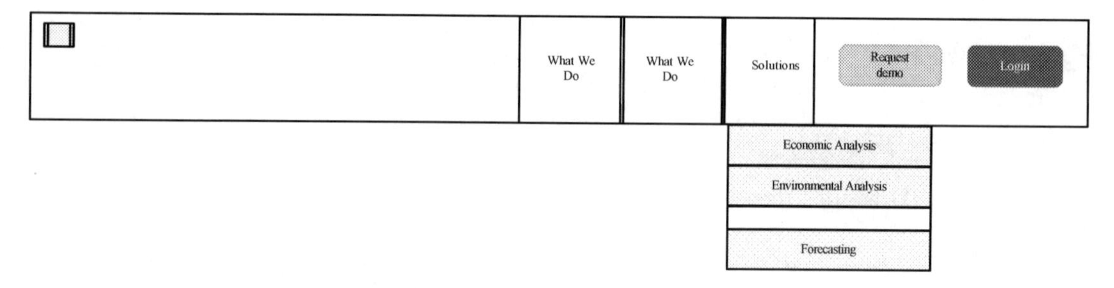

Figure 10-6. *Final navigation bar*

Footer Navigation Bar

This section constructs a footer that contains navigation items that direct the user to some link. It can reuse prespecified navigation items and/or include new items. Listing 10-14 constructs navigation items.

Listing 10-14. Constructs Navigation Items

```
about = html.Div(
    [
    html.H4("About", style={"color":"white"})
    ]
)
about_item1 = dbc.NavItem(dbc.NavLink("Our Company",
                                      href = "#",
                                      style = {"color":"white"}
                                     )
                          )
about_item2 = dbc.NavItem(dbc.NavLink("Careers",
                                      href = "#",
                                      style = {"color":"white"}
                                     )
                          )
about_item3 = dbc.NavItem(dbc.NavLink("FAQ",
                                      href = "#",
                                      style = {"color":"white"}
                                     )
                          )
about_fullitems = html.Div(
    [
        about,
        about_item1,
        about_item2,
        about_item3
    ]
)
legal = html.Div(
    [
        html.H4("Legal", style={"color":"white"})
    ])
```

```
legal_item1 = dbc.NavItem(dbc.NavLink("Term of use",
                                      href = "#",
                                      style = {"color":"white"}
                                     )
                         )
legal_item2 = dbc.NavItem(dbc.NavLink("Anti-Spam Policy",
                                      href = "#",
                                      style = {"color":"white"}
                                     )
                         )
legal_item3 = dbc.NavItem(dbc.NavLink("Cookie Policy",
                                      href = "#",
                                      style = {"color":"white"}
                                     )
                         )
legal_fullitems = html.Div([legal,
                            legal_item1,
                            legal_item2,
                            legal_item3])

resource = html.Div(
    [
        html.H4("Resource", style={"color":"white"})
    ])
resource_item1 = dbc.NavItem(dbc.NavLink("Information & Webinars",
                                         href = "#",
                                         style = {"color":"white"}
                                        )
                            )
resource_item2 = dbc.NavItem(dbc.NavLink("News & Insights",
                                         href = "#",
                                         style = {"color":"white"}
                                        )
                            )
```

```
resource_item3 = dbc.NavItem(dbc.NavLink("Learning Portal",
                                          href = "#",
                                          style = {"color":"white"}
                                         )
                             )
resource_fullitems = html.Div(
    [
        resource,
        resource_item1,
        resource_item2,
        resource_item3
    ]
)
solutions_init = html.Div(
    [
        html.H4("Solutions", style={"color":"white"})
    ])
solutions_item1 = dbc.NavItem(dbc.NavLink("Economic & Financial Markets
Research",
                                          href = "#",
                                          style = {"color":"white"}
                                         )
                             )
solutions_item2 = dbc.NavItem(dbc.NavLink("Environment Analysis",
                                          href = "#",
                                          style = {"color":"white"}
                                         )
                             )
solutions_item3 = dbc.NavItem(dbc.NavLink("Forecast",
                                          href = "#",
                                          style = {"color":"white"}
                                         )
                             )
```

```
solutions_fullitems = html.Div(
    [
        solutions_init,
        solutions_item1,
        solutions_item2,
        solutions_item3
    ]
)
```

Listing 10-15 groups the navigation items for a footer navigation bar.

Listing 10-15. Group Navigation Items for Footer Navigation Bar

```
footer_nav = dbc.NavbarSimple(
    children=[about_fullitems,
              solutions_fullitems,
              resource_fullitems,
              legal_fullitems],
    brand="",
    brand_href="#",
    sticky="bottom",
    className="mb-5",
    color="dark",
    dark=True
)
```

It is important to specify copyright reserved as a mechanism for protecting intellectual property. Listing 10-16 specifies the copyright section (see Figure 10-7).

Listing 10-16. Copyright

```
copyrights = dbc.Card([
    dbc.CardBody([
        dbc.Row([
            dbc.Col([], width = 5),
            dbc.Col([
                html.P("Copyright © 2021 WorldViewer All rights reserved.",
                       style={"color":"dark"}
```

```
                    )
                ]
                )
            ]
            )
        ]
        )
],
    color="light")
```

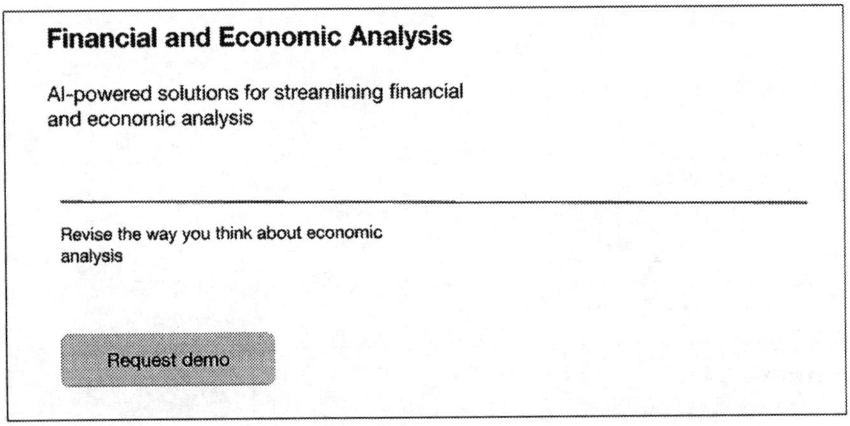

Figure 10-7. *Final footer navigation bar*

Banner

Listing 10-17 constructs a banner that proceeds to the top navigation bar (see Figure 10-8).

Listing 10-17. Banner

```
banner1 = html.Img(src = app.get_asset_url("banner1.png"))
home_jumbotron = dbc.Jumbotron(
    [
        dbc.Row([
            dbc.Col([],width=2),
            dbc.Col([
                html.H4("Financial and Economic Analysis",
                        className="display-4"),
```

```
                html.H3("AI-powered solutions for streamlining financial
                and economic analysis",
                        className = "lead",
                       ),
                html.Hr(className = "my-2"),
                html.H3("Revise the way you think about economic analysis",
                       ),
                html.P(dbc.Button("Request demo",
                                  color = "dark",
                                  outline = True),
                       className = "lead")
            ]
            )
        ]
        )
    ]
)
```

Figure 10-8. *Banner*

Listing 10-18 specifies the content and layout of the app.

Listing 10-18. Specify Content and App Layout

```
content = html.Div(id = "page-content")
app.layout = html.Div(
    [
        default, home_jumbotron, app_features,
        value_proposition, subscribe_card, footer_nav,
        copyrights, content]
)
```

Callback to Collapse the Navigation for Small Screens

Listing 10-19 constructs a callback that collapses on small screens.

Listing 10-19. Callback to Collapse the Navigation for Small Screens

```
def toggle_navbar_collapse(n, is_open):
    if n:
        return not is_open
    return is_open
for i in [1, 2, 3]:
    app.callback(
        Output(f"navbar-collapse{i}", "is_open"),
        [Input(f"navbar-toggler{i}", "n_clicks")],
        [State(f"navbar-collapse{i}", "is_open")],
    )(toggle_navbar_collapse)
```

Listing 10-20 runs the Dash app (see Figure 10-9).

Listing 10-20. Run Dash App

```
app.run_server(mode="external",
               dev_tools_ui=False,
               dev_tools_props_check=False)
```

Home Page

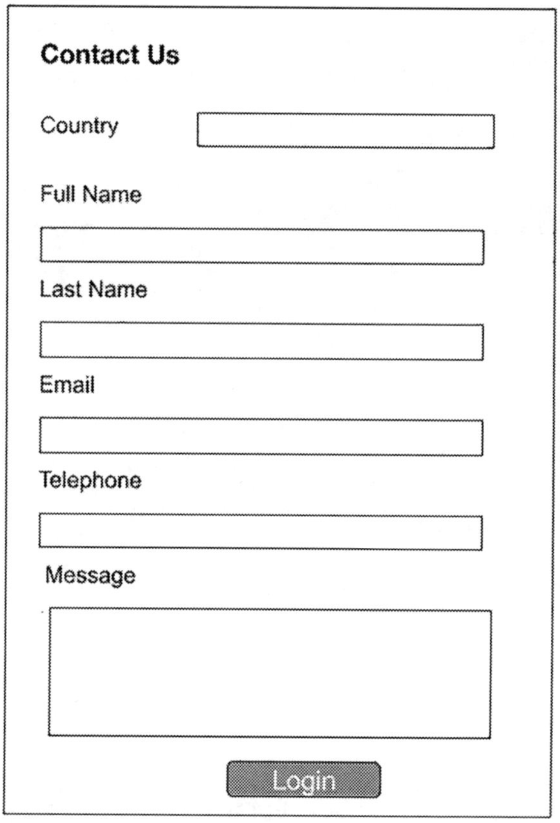

Figure 10-9. *Home page*

Contact Us

A contact page enables a user to communicate a message to an organization. It also allows users to specify their contact details, so the organization responds accordingly.

Listing 10-21 initializes the app.

Listing 10-21. Initializing App

```
font_awesome = "https://use.fontawesome.com/releases/v5.8.1/css/all.css"
app = JupyterDash(external_stylesheets=[dbc.themes.MATERIA, font_awesome],
                  meta_tags=[{"charset":"utf-8",
                             "name": "viewport",
                             "content": "width=device-width, initial-
                             scale=1"}])
```

Listing 10-22 specifies the style of the page using the CSS format.

Listing 10-22. Content Styling

```
CONTENT_STYLE = {
    "margin-left": "16rem",
    "margin-right": "0.5rem",
    "padding": "0.5rem 0.5rem",
    "background-color": "white"
}
```

Listing 10-23 specifies individual components (i.e., `country_list`, `full_name`, `last_name`, `email_input`, `phone_number`, and `textareas`). Inputs contain placeholders, which is the text that informs a user about what to do.

Listing 10-23. Form Creation

```
country_list = dbc.FormGroup(
    [
        dbc.Label("Country", html_for="example-password-row", width=3),
        dbc.Col(
            dbc.Select(
                id="select",
                options=[
                    {"label": "South Africa", "value": "1"},
                    {"label": "United States", "value": "2"},
                    {"label": "Israel", "value": "3"},
                ],
            )
```

```
        ),
    ],
    row=True)
email_input = dbc.FormGroup(
    [
        dbc.Label("Email",
                  html_for = "email-input",
                  width=3),
        dbc.Col(
            dbc.Input(
                type="email",
                id = "contact-password-input",
                placeholder = "Enter email"
            ),
            width=10,
        ),
    ],
    row=True,
)

full_name = dbc.FormGroup(
    [
        dbc.Label("Full Name",
                  html_for = "fullname-input",
                  width=3),
        dbc.Col(
            dbc.Input(
                id = "contact-full-name",
                placeholder="Enter full name",
                bs_size="md",
                className="mb-3"),
            width=10,
        ),
    ],
    row=True,
)
```

```
last_name = dbc.FormGroup(
    [
        dbc.Label("Last Name",
                  html_for="example-password-row",
                  width = 3),
        dbc.Col(
            dbc.Input(
                id = "concat-last-name",
                placeholder = "Enter last name",
                bs_size="md",
                className="mb-3"),
            width=10,
        ),
    ],
    row=True,
)
phone_number = dbc.FormGroup(
    [
        dbc.Label("Phone",
                  html_for = "contact-phone-number",
                  width = 3),
        dbc.Col(
            dbc.Input(
                id = "contact-phone-number",
                type = "number",
                placeholder = "Enter phone ",
                bs_size = "md",
                className = "mb-3"),
            width = 10,
        ),
    ],
    row=True,
)
```

```
buttons = html.Div(
    [
        dbc.Button("Send",
                   id = "conctact-send-button",
                   color = "primary",
                   block = True),
    ]
)
textareas = dbc.FormGroup(
    [
        dbc.Label("Message",
                  html_for="contact-textarea",
                  width = 3),
        dbc.Col(
            dbc.Textarea(
                id = "contact-textarea",
                bs_size = "lg",
                placeholder = "Type message"
            ),
            width=10,
        ),
    ],
    row=True,
)
```

Listing 10-24 groups components created in Listing 10-23 (i.e., country_list, full_name, last_name, email_input, phone_number, and textareas).

Listing 10-24. Group Contact Us Components

```
header = html.H2("Contact Us")
form = dbc.Form([header,
                html.Br(),
                country_list,
                full_name,
                last_name,
```

```
                email_input,
                phone_number,
                textareas,
                html.Br(),
                buttons])
card_form = dbc.Card(dbc.CardBody([form],
                    ),style={"margin-left": "22rem",
                            "margin-right": "22rem",
                            "margin-top": "8rem",
                              "margin-bottom": "3rem",
                            "padding": "0.5rem 0.5rem",
                            "background-color": "white"})
copyrights = dbc.Card([
    dbc.CardBody([
        dbc.Row([
            dbc.Col([],width=5),
                        dbc.Col([
                            html.P("Copyright © 2021 Worldviewer All
                            rights reserved.",
                                  style = {"color":"dark"}
                                 )
                        ]
                        )
        ]
        )
    ]
    )
],
    color="light")
contactus_jumbotron = dbc.Jumbotron(
    [
        card_form,

    ]
)
```

Listing 10-25 specifies the content and layout of the app.

Listing 10-25. Specifying the Content and App Layout

```
content = html.Div(id = "page-content")
app.layout = html.Div(
    [
        contactus_jumbotron,
        copyrights,
        content
    ]
)
```

Listing 10-26 runs the app (see Figure 10-10).

Listing 10-26. Run App

```
app.run_server(mode = "external",
               dev_tools_ui = False,
               dev_tools_props_check = False)
```

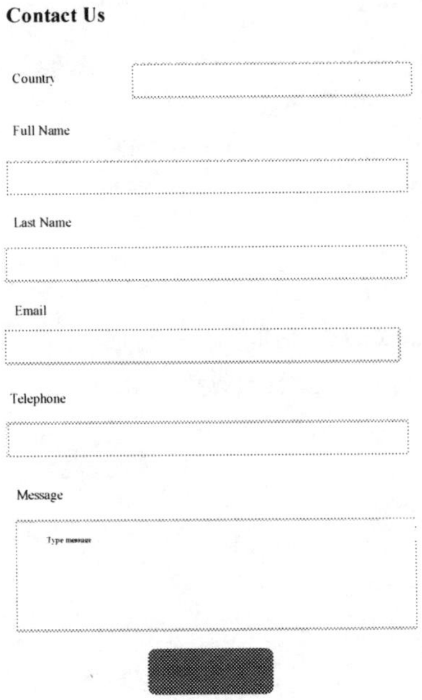

Figure 10-10. *Contact us*

Billing/Checkout

A billing/checkout page enables users to provide their bank card details and payment information to process a transaction through a getaway. This is important if a site sells products.

Listing 10-27 constructs components for the billing/checkout page (i.e., `header_banking_details`, `name_on_card`, `credit_card_number`, `cvv_input`, and `buttons`). Each input component contains a placeholder that informs users of the information they should enter.

Listing 10-27. Billing/Checkout Components

```
name_on_card = dbc.FormGroup(
    [
        dbc.Label("Name on card",
                  html_for = "bank-card-name",
                  width = 3),
        dbc.Col(
            dbc.Input(
                id = "bank-card-name",
                placeholder = "Enter name on card",
                bs_size = "md",
                className = "mb-3"),
            width=10,
        ),
    ],
    row = True,
)
credit_card_number = dbc.FormGroup(
    [
        dbc.Label("Card number",
                  html_for = "bank-card-number",
                  width = 3),
        dbc.Col(
            dbc.Input(
                id = "bank-card-number",
                placeholder = "Enter credit card number",
                bs_size = "md",
                className = "mb-3"),
            width = 10,
        ),
    ],
    row  =True,
)
```

```
cvv_input = dbc.FormGroup(
    [
        dbc.Label("CVV",
                  html_for = "bank-cvv",
                  width = 3),
        dbc.Col(
            dbc.Input(
                id = "bank-cvv",
                placeholder = "MM/YY",
                bs_size = "md",
                className = "mb-3"),
            width = 10,
        ),
    ],
    row = True,
)
check_button = html.Div(
    [
        dbc.Button("Check Out",
                   id = "check-out-button",
                   color = "primary",
                   block = True),
    ]

)
radioitems = dbc.FormGroup(
    [
        dbc.Label("Payment option"),
        dbc.RadioItems(
            options=[
                {"label": "Credit card", "value": 1},
                {"label": "Debit card", "value": 2},
            ],
            value = 1,
            id = "radioitems-input",
        ),
    ]
```

```
)
header_billing = html.H2("Billing address")
header_banking_details = html.H2("Banking details")
baking_details = dbc.Card([
    dbc.CardBody([])])
forgot_password = html.Label(['',
                                   html.A('Forgot Password',
                                          href = '#')])
form = dbc.Form(
    [
        header_billing,
        radioitems,
    ]
)
```

Listing 10-28 groups components for the billing/checkout form (i.e., header_banking_details, name_on_card, credit_card_number, cvv_input, and buttons).

Listing 10-28. Group Billing/Checkout Components

```
form2 = dbc.Form(
    [
        header_banking_details,
        name_on_card,
        credit_card_number,
        cvv_input,
        html.Br(),
        check_button
    ]
)
card_form = dbc.Card([
    dbc.CardBody([form,
                  html.Br(),
                  form2]),
    style = {"margin-left": "22rem",
             "margin-right": "22rem",
             "margin-top": "8rem",
```

```
            "margin-bottom": "3rem",
            "padding": "0.5rem 0.5rem",
            "background-color": "white"}])
copyrights = dbc.Card([
    dbc.CardBody([
        dbc.Row([
            dbc.Col([], width = 5),
            dbc.Col([
                html.P("Copyright © 2021 Worldviewer All rights reserved.",
                       style = {"color":"dark"}
                      )
            ]
            )
        ]
        )
    ]
    )
],
    color="light")
checkout_jumbotron = dbc.Jumbotron(
    [
        card_form,

    ]
)
```

Listing 10-29 specifies the content and Dash web app layout of the app.

Listing 10-29. Specifying the Content and App Layout

```
content = html.Div(id="page-content")
app.layout = html.Div(
    [
        checkout_jumbotron,
        copyrights,
        content
    ]
)
```

Listing 10-30 runs the app (see Figure 10-11).

Listing 10-30. Run the App

```
app.run_server(mode = "external",
               dev_tools_ui = False,
               dev_tools_props_check = False)
```

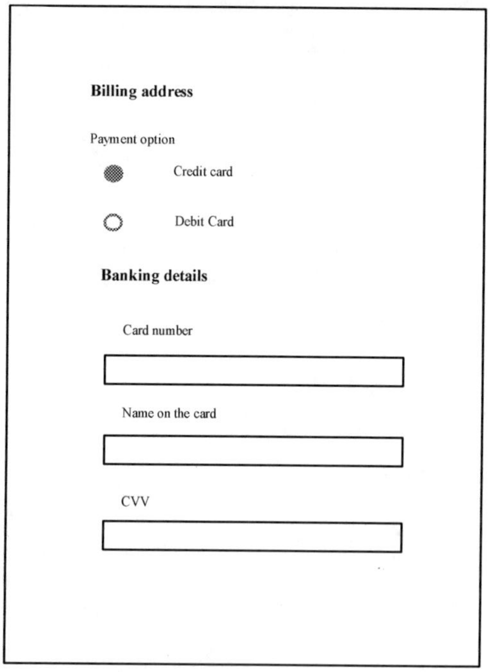

Figure 10-11. *Checkout*

Conclusion

This chapter acquainted you with the basics of developing a full website based on Dash Bootstrap Components, Dash HTML Components, and some basic CSS styling functionality. Successively, you may include JavaScript and other languages, including other web APIs, to develop ideal websites. An ideal place to get started is W3Schools.

CHAPTER 11

Integrating a Machine Learning Algorithm into a Web App

This chapter introduces an approach to integrate machine learning models. It overviews linear regression, including ways to preprocess data and generate predictions. The chapter concludes by demonstrating a technique to implement machine learning in web apps.

An Introduction to Linear Regression

Linear regression is a machine learning model that predicts a continuous dependent feature (also known as a *dependent variable*) based on a set of independent features. A variable or feature represents any continuous process.

Equation 11-1 is a linear function.

$$\hat{y} = \beta_0 + \beta_1 X_1 + \varepsilon_i \quad \text{(Equation 11-1)}$$

$\hat{y}$ constitutes the values of a dependent feature that the linear regression algorithm estimates. β_0 constitutes the intercept (the mean value of the dependent feature holds an independent feature constant). β_1 constitutes the slope. X_1 constitutes the independent feature and makes up the difference between the actual and predicted values of a dependent feature.

Figure 11-1 illustrates how the algorithm works.

T. C. Nokeri, *Web App Development and Real-Time Web Analytics with Python*,
https://doi.org/10.1007/978-1-4842-7783-6_11

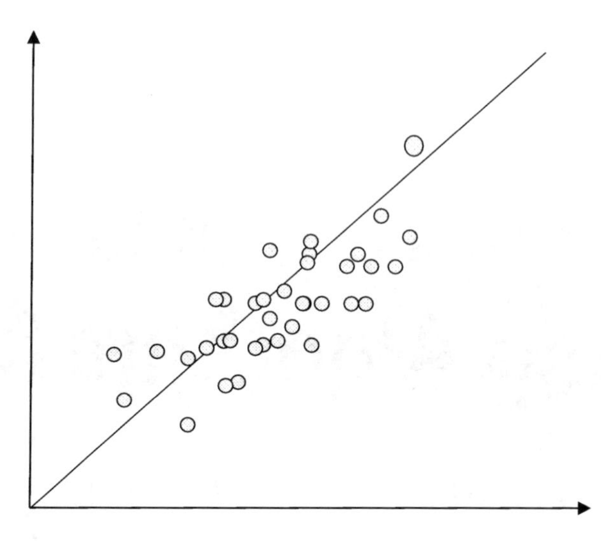

Figure 11-1. *Linear function*

Figure 11-1 shows the values scattered across the graph. A straight line cuts through the values.

An Introduction to sklearn

sklearn is the most prevalent Python library for solving machine learning problems. This chapter preprocesses the data and trains the algorithm using sklearn. First, ensure that you have the sklearn library installed in your environment. To install the sklearn library in a Python environment, use `pip install sklearn`. To install the library in a conda environment, use `conda install -c anaconda scikit-learn`.

First, import data into a pandas DataFrame. This chapter applies the data used in Chapters 1 and 2 (see Table 11-1).

Table 11-1. DataFrame

	gdp_by_exp	cpi	m3	spot_crude_oil	rand
DATE					
2009-01-01	−1.718249	71.178127	13.831098	41.74	9.3000
2009-04-01	−2.801610	73.249160	9.774203	49.79	9.3705
2009-07-01	−2.963243	74.448179	5.931918	64.09	7.7356
2009-10-01	−2.881582	74.884186	3.194678	75.82	7.7040
2010-01-01	0.286515	75.320193	0.961220	78.22	7.3613

Preprocessing

After importing the data, the next step involves preprocessing the data. First, you impute the data (substituting missing values in a feature by some values). This chapter imputes the data using the mean value implementation strategy. A mean value is the arithmetic average of values in a feature. Listing 11-1 imputes the data by implementing the mean-value imputation strategy.

Listing 11-1. Impute Data using the Mean Value Imputation Strategy

```
df["gdp_by_exp"] = df["gdp_by_exp"].fillna(df["gdp_by_exp"].mean())
df["cpi"] = df["cpi"].fillna(df["cpi"].mean())
df["m3"] = df["m3"].fillna(df["m3"].mean())
df["spot_crude_oil"] = df["spot_crude_oil"].fillna(df["spot_crude_oil"].mean())
df["rand"] = df["rand"].fillna(df["rand"].mean())
```

After imputation, assign the features. In this example, GDP by expenditure, consumer price index, and money supply are independent features, and rand is a dependent feature.

Listing 11-2 assigns x (independent features) and y (dependent feature).

Listing 11-2. Assign Independent and Dependent Feature(s)

```
x = df.iloc[::,0:4]
y = df.iloc[::,-1]
```

Splitting Data into Training and Test Data

Let's split the data into training and test data. The algorithm learns the training data and predict the data. Listing 11-3 splits data into training and test data using the `train_test_split()` method from the `model_selection` function in the sklearn library.

Listing 11-3. Splitting Data into Training and Test Data

```
from sklearn.model_selection import train_test_split
x_train, x_test, y_train, y_test = train_test_split(x, y, test_size=0.2)
```

Standardization

Let's standardize the data using the `StandardScaler()` method. Standardization transforms by centering it so that the mean value is 0 and the standard deviation is 1. Listing 11-4 standardizes the data by implementing the `StandardScaler()` method from the `preprocessing` function in sklearn.

Listing 11-4. Standardizing Data

```
from sklearn.preprocessing import StandardScaler
scaler = StandardScaler()
x_train = scaler.fit_transform(x_train)
x_test = scaler.transform(x_test)
```

Training an Algorithm

Next, let's train the linear regression algorithm. Call the `LinearRegression()` method from the `linear_model` function in the sklearn library. Listing 11-5 trains the algorithm.

Listing 11-5. Training a Linear Regression Algorithm

```
from sklearn.linear_model import LinearRegression
linear_model = LinearRegression()
linear_model.fit(x_train, y_train)
```

Predictions

After training, generate values of the dependent feature that the algorithm predicts. Listing 11-6 generates values of the dependent features the linear regression algorithm predicts by implementing the `predict()` method from sklearn, and then tabulates the values by implementing pandas (see Table 11-2).

Listing 11-6. Predict Values of the Dependent Features

```
y_pred_dates = df[:10].index
y_pred = pd.DataFrame(linear_model.predict(x_test), columns = ["Predicted rand"])
y_pred.index = y_pred_dates
y_pred
```

Table 11-2. *Actual and Predicted Values of Rand*

	Predicted rand
DATE	
2009-01-01	13.901479
2009-04-01	7.854847
2009-07-01	8.660124
2009-10-01	15.219995
2010-01-01	13.793463
2010-04-01	14.648314
2010-07-01	7.854163
2010-10-01	19.729906
2011-01-01	13.516257
2011-04-01	12.587248

Integrating an Algorithm to a Web App

Earlier, you saw a way to train a machine learning algorithm using the scikit-learn library. This section develops an app from scratch and integrates the algorithm into the app, which predicts future instances based on stocks that a user selects.

Listing 11-7 extracts the data by implementing the `read_excel()` method from the pandas library. Subsequently, it captures the symbols in the list to serve as options in the search drop-down menu in the Dash app by implementing the `append()` method from the pandas library.

Table 11-3 highlights the data contained in the Microsoft Excel file document.

Listing 11-7. Extracting Country Search Options

```
stock_ticker = pd.read_excel(r"filepath\allassets.xlsx")
stock_ticker = stock_ticker.set_index("Symbol")
options = []
for tic in stock_ticker.index:
    options.append({"label":"{} {}".format(tic,stock_ticker.loc[tic]
    ["Name"]), "value":tic})
```

Table 11-3. *Stock Listing*

	Symbol	Name
0	MMM	3M Company
1	AOS	A.O. Smith Corp
2	ABT	Abbott Laboratories
3	ABBV	AbbVie Inc.
4	ABMD	ABIOMED Inc
...	...	...
501	ZBRA	Zebra Technologies
502	ZBH	Zimmer Biomet Holdings
503	ZION	Zions Bancorp
504	ZTS	Zoetis
505	Tata Motors Limited	Tata Motors Limited

Listing 11-8 extracts a CSS file from Bootstrap, a widely used CSS provider.

Listing 11-8. Loading an External CSS File

```
get_bootstrap_css = "https://cdn.jsdelivr.net/npm/bootstrap@5.0.0-beta2/
dist/css/bootstrap.min.css"
```

Listing 11-9 extracts the icon library from Bootstrap.

Listing 11-9. Loading an Icons Library

```
get_bootsrap_icon = "https://cdn.jsdelivr.net/npm/bootstrap-icons@1.4.0/
font/bootstrap-icons.css"
```

Initializing a Web App

Listing 11-10 initializes the app by implementing the JupyterDash library.

Listing 11-10. Initializing a Web App

```
app = JupyterDash(external_stylesheets=[get_bootstrap_css, get_bootsrap_icon],
                  meta_tags=[{"name" : "viewport",
                              "charset" :"utf-8",
                              "content" : "width=device-width,initial-
                              scale=1, shrink-to-fit=no"}])
```

Navigation Bars

Listing 11-11 specifies navigation bars. The code is similar to the one contained in Chapter 8.

Listing 11-11. Specify Navigation Bars

```
alerts_notif = dbc.Row(
    [
        dbc.Col(
            [dbc.NavLink(className="bi bi-bell",
                         href = "/page-2/1",
                        style = {"font-size" : "20px", "color" : "gray"}),],
            width = "auto"
        ),
    ],
    no_gutters = True,
    className = "ml-auto flex-nowrap mt-3 mt-md-0",
    align = "right",
    id = "responsivemenu-3-collapse"
)
messages = dbc.Row(
    [
        dbc.Col(
            dbc.NavLink(className = "bi bi-envelope",
                        href = "/page-3/1",
                        style = {"font-size" : "20px", "color" : "gray"}),
            width="auto"
        ),
    ],
    no_gutters = True,
    className = "ml-auto flex-nowrap mt-3 mt-md-0",
    align = "right",
     style = {"font-size" : "16px"},
    id = "responsivemenu-4-collapse"
)
profile = dbc.DropdownMenu(
    children=[
        dbc.DropdownMenuItem("Edit Profile",
                             href = "/page-4/1"),
        dbc.DropdownMenuItem("Privacy & Safety",
```

```
                                  href = "/page-4/2"),
            dbc.DropdownMenuItem("Account Settings",
                                  href = "/page-4/3"),
            dbc.DropdownMenuItem("Billing",
                                  href = "/page-4/4"),
            dbc.DropdownMenuItem(divider = True),
            dbc.DropdownMenuItem("Sign Out",
                                  href = "/page-4/5"),
        ],
        nav = True,
        in_navbar = True,
        className = "bi bi-person",
        direction = "left",
        style = {"font-size" : "20px"},
        id = "responsivemenu-5-collapse"
)
navbar = dbc.Navbar(
        [
            dbc.Col([],width=2),
            dbc.Col([
                dbc.Button(id = "toggle-button",
                           n_clicks = 0,
                           children = "",
                           outline = True,
                           className = "navbar-toggler-icon")],
                width = 1),
            dbc.Col([],
                    width = 3),
            dbc.Col([messages],
                    width = "auto"),
            dbc.Col([alerts_notif],
                    width = "auto"),
            dbc.Col([], width = 2),
            dbc.Col([profile],
                    width = 2)],
```

```
    color = "white",
    style={"margin-right" : "0rem",
            "margin-top" : "0rem",
            "margin-bottom" : "0.5rem",
            "padding" : "1rem 0rem"})
RESPONSIVE_MENU_STYLE = {
    "position" : "fixed",
    "top" : 0,
    "left" : 0,
    "bottom" : 0,
    "width" : "14rem",
    "height" : "100%",
    "margin-top" : "0rem",
    "margin-bottom" : "0rem",
    "z-index" : 1,
    "overflow-x" : "hidden",
    "transition" : "all 0.5s",
    "padding" : "0.5rem 1rem"
}
RESPONSIVE_MENU_HIDEN = {
    "position" : "fixed",
    "top" : 0,
    "left" : "-16rem",
    "bottom" : 0,
    "width" : "14rem",
    "height" : "100%",
    "z-index" : 1,
    "overflow-x" : "hidden",
    "transition" : "all 0.5s",
    "padding" : "0rem 0rem",
}
APP_CONTENT_STYLE = {
    "transition" : "margin-left .5s",
    "margin-left" : "14.5rem",
    "margin-right" : "0.5rem",
```

```
    "margin-bottom" : "0.5rem",
    "padding" : "0rem 0rem"
}
APP_CONTENT_STYLE1 = {
    "transition" : "margin-left .5s",
    "margin-left" : "0.5rem",
    "margin-right" : "0.5rem",
    "margin-bottom" : "0.5rem",
    "padding" : "0rem 0rem"
}
RESPONSIVE_RESPONSIVEMENU_1 = [
    html.Li(
        dbc.Row(
            [
                dbc.Col(dbc.NavLink("Forecast Indicators",
                        href = "/page-1/1",
                        style = {"color" : "#616161"})),
            ],
            className = "my-1",
        ),
        style = {"cursor" : "pointer"},
        id = "responsivemenu-1",
    )
]

RESPONSIVE_SIDE_NAVIGATION_BAR = html.Div(
    [
        dbc.Card(
            [
                dbc.CardBody(
                    [
                        html.H4("WorldViewer",
                                className = "btn btn-outline-primary"),
                        html.Hr(),
```

```
                        html.P(
                            "",
                            className = "lead"),
                        dbc.Nav(
                            RESPONSIVE_RESPONSIVEMENU_1,
                            vertical = True)
                    ]
                )
            ],
            id = "responsivesidebar")
    ]
)
```

Search Functionality

Listing 11-12 develops search components by implementing the `Dropdown()` method from the `dash_core_components` library. The two components are for searching independent variables and the dependent variable. It also holds a button that unhides results upon clicking.

Listing 11-12. Search Functionality

```
INPUT_CARD = dbc.Card([
    dbc.CardBody([
        dbc.Row([
            dbc.Col([
                dbc.FormGroup([
                    dbc.Label("Select independent variables (over 3
                    stocks)"),
                    dcc.Dropdown(id = "independent-variable-input",
                                 className = "border-bottom",
                                 options = options,
                                 multi = True,
                                 value = [],
                                 placeholder = "Search stocks (independent
                                 variables)")])],
                width = 6),
```

```
        dbc.Col([
            dbc.FormGroup([
                dbc.Label("Select the dependent variable"),
                dcc.Dropdown(id = "dependent-variable-input",
                             className = "border-bottom",
                             options = options,
                             multi = False,
                             placeholder = "Search stock (dependent
                             variable)")])],
            width = 6)]),
    html.Br(),
    dbc.Row([
        dbc.Col([], width=5),
        dbc.Col([
            dbc.Button("Show results",
                       id = "worldviewer-submit-button",
                       color = "primary")],
            width = 4)])])])
```

Containing Interactive Tables for Results

Listing 11-13 contains interactive tables for results. The tables include a descriptive statistics table, a correlation matrix, a prediction table, an intercept and coefficients table, and a model evaluation table.

Listing 11-13. Containing Interactive Tables for Results

```
OUTPUT_CARD = html.Div([dbc.Card([
    dbc.CardBody([
            html.H4("Statistical summary",
                    className="card-title",
                    style={"text-align": "center"})]),],
        style={"width": "auto","background-color": "#FAFAFA"}),
    html.Br(),
    dbc.Row([dbc.Col([html.Div(id="descriptive-statistics")])]),
    html.Br(),
```

```
dbc.Card([
    dbc.CardBody([
        html.H4("Correlation",
                className="card-title",
                style={"text-align": "center"})]),],
    style={"width": "auto","background-color": "#FAFAFA"},),
html.Br(),
dbc.Row([dbc.Col([html.Div(id="pearson-correlation")])]),
html.Br(),
dbc.Card([
    dbc.CardBody([
        html.H4("Prediction",
                className="card-title",
                style={"text-align": "center"})]),],
    style={"width": "auto","background-color": "#FAFAFA"},),
html.Br(),
dbc.Row([dbc.Col([html.Div(id="linear-model-predictions")])]),
html.Br(),
dbc.Card([
    dbc.CardBody([
        html.H4("Coefficients and intercept",
                className="card-title",
                style={"text-align": "center"})]),],
    style={"width": "auto","background-color": "#FAFAFA"},),
html.Br(),
dbc.Row([dbc.Col([html.Div(id="linear-model-intercept-and-
coefficients")])]),
html.Br(),
dbc.Card([
    dbc.CardBody([
        html.H4("Model evaluation",
                className="card-title",
                style={"text-align": "center"})]),],
    style={"width": "auto","background-color": "#FAFAFA"},),
```

```
html.Br(),
dbc.Row([dbc.Col([html.Div(id="linear-model-evaluation")])]),
])
```

Specifying the App Layout and Callbacks for Responsive Side Menus and URL Routing

Listing 11-14 specifies the app layout and callbacks for responsive side menus and URL routing. The code is similar to the one specified in Chapter 8. Moreover, there are a few changes. For instance, the app layout comprises the following components: `dcc.Store(id='independent-variables-values')` which holds data relating to independent variables and `dcc.Store(id='dependent-variable-value')` which holds data relating to the dependent variable, so as `dcc.Location(id = "url")`. This approach enables the app to temporarily store the data on the browser rather than storing it in some database.

Listing 11-14. Specifying the App Layout and Callbacks for Responsive Side Menus and URL Routing

```
WORDVIEWER_SUMMARY = html.Div([
    dbc.Row([dbc.Col([OUTPUT_CARD])])],
    id="collapse-worldviewer-page")
WORDVIEWER_LAYOUT = html.Div([
    html.Br(),
    dbc.Row([
        dbc.Col([INPUT_CARD],width=12)]),
    html.Br(),
    dbc.Row([
        dbc.Col([],width=5),
        dbc.Col([])]),
    dbc.Collapse([WORDVIEWER_SUMMARY],
                 id="collapse-worldviewer-menu")])
content = html.Div(
    id = "app-content",
    style=APP_CONTENT_STYLE)
```

```
app.layout = html.Div(
    [
        dcc.Store(id = "responsive-sidebar-click"),
        dcc.Store(id='independent-variables-values'),
        dcc.Store(id='dependent-variable-value'),
        dcc.Location(id = "url"),
        navbar,
        RESPONSIVE_SIDE_NAVIGATION_BAR,
        content
    ]
)
@app.callback(
    [
        Output("responsivesidebar", "style"),
        Output("app-content", "style"),
        Output("responsive-sidebar-click", "data"),
    ],

    [Input("toggle-button", "n_clicks")],
    [
        State("responsive-sidebar-click", "data"),
    ]
)
def toggle_responsivesidebar(n, nclick):
    if n:
        if nclick == "SHOW":
            RESPONSIVE_MENU_style = RESPONSIVE_MENU_HIDEN
            APP_CONTENT_style = APP_CONTENT_STYLE1
            NO_OF_CURRENT_CLICKS = "HIDDEN"
        else:
            RESPONSIVE_MENU_style = RESPONSIVE_MENU_STYLE
            APP_CONTENT_style = APP_CONTENT_STYLE
            NO_OF_CURRENT_CLICKS = "SHOW"
```

```
    else:
        RESPONSIVE_MENU_style = RESPONSIVE_MENU_STYLE
        APP_CONTENT_style = APP_CONTENT_STYLE
        NO_OF_CURRENT_CLICKS = "SHOW"
    return RESPONSIVE_MENU_style, APP_CONTENT_style, NO_OF_CURRENT_CLICKS
def toggle_collapse(n, is_open):
    if n:
        return not is_open
    return is_open
def set_navitem_class(is_open):
    if is_open:
        return "open"
    return ""
path_name_map = {"/": WORDVIEWER_LAYOUT,
                 "/page-1/1": WORDVIEWER_LAYOUT,
                 "/page-2/1": "Forecast Indicators",
                 "/page-3/1": "Profile",
                 "/page-4/1": "Edit Profile",
                 "/page-4/2": "Privacy & Safety",
                 "/page-4/3": "Account Settings",
                 "/page-4/4": "Billing",
                 "/page-4/5": "Sign Out"}
for i in range(0, 4):
    app.callback(
        Output(f"responsivemenu-{i}-collapse", "is_open"),
        [Input(f"responsivemenu-{i}", "n_clicks")],
        [State(f"responsivemenu-{i}-collapse", "is_open")],
    )(toggle_collapse)
    app.callback(
        Output(f"responsivemenu-{i}", "className"),
        [Input(f"responsivemenu-{i}-collapse", "is_open")],
    )(set_navitem_class)
@app.callback(Output("app-content", "children"), [Input("url", "pathname")])
def render_page_content(pathname):
    return html.P(path_name_map[pathname])
```

Listing 11-15 specifies a callback function for unhiding content upon the user clicking the "Show results" button.

Listing 11-15. Specifying a Callback Function for an Unhiding Content

```
@app.callback(Output("collapse-worldviewer-menu", "is_open"),
              [Input("worldviewer-submit-button", "n_clicks")])
def toggle_collapse_worldviewer_menu(n):
    return n
```

Specifying a Callback to Load Independent Variables Values

Listing 11-16 specifies a callback for loading independent variable values. It specifies the `start` date and end date and then collects the data relating to the selected stock using the `pandas-datareader` library. It concludes by loading the data using the `to_json()` method.

Listing 11-16. Specifying a Callback to Load Independent Variables Values

```
@app.callback(Output("independent-variables-values", "data"),
              [Input("independent-variable-input", "value")])
def independepnt_variables_dataset(stock_ticker):
    start = "2018-01-01"
    end = "2021-09-01"
    df = web.get_data_yahoo(stock_ticker, start = start, end = end)
    ind_variable_data = pd.DataFrame(df.Close)
    ind_variable_data = ind_variable_data.dropna()
    return ind_variable_data.to_json(date_format = "iso", orient = "split")
```

Specifying a Callback for Loading the Dependent Variable Values

Listing 11-17 specifies a callback for loading the dependent variable values. It specifies the `start` date and end date and then collects the data relating to the selected stock using the `pandas-datareader` library. It concludes by loading the data using the `to_json()` method.

Listing 11-17. Specifying a Callback for Loading the Dependent Variable Values

```
@app.callback(Output("dependent-variable-value", "data"),
              [Input("dependent-variable-input", "value")])
def dependent_variable_dataset(stock_ticker):
    start = "2018-01-01"
    end = "2021-09-01"
    trace = []
    for tic in stock_ticker:
        df = web.get_data_yahoo(stock_ticker, start = start, end = end)
        ind_variable_data = pd.DataFrame(df.Close)
        ind_variable_data = ind_variable_data.dropna()
        labels = stock_ticker
        ind_variable_data.columns = [labels]
        return ind_variable_data.to_json(date_format = "iso", orient = "split")
```

Specifying a Callback for Descriptive Statistics

Listing 11-18 specifies a callback for an algorithm's intercept and coefficients. First, it loads the independent variables and the dependent variable by implementing the read_json() method. After that, it discerns descriptive statistics containing the mean value, median value, and standard deviation and then passes the values into a table.

Listing 11-18. Specifying a Callback for Descriptive Statistics

```
@app.callback(Output("descriptive-statistics", "children"),
              [Input("independent-variables-values", "data")],
              [State("dependent-variable-value", "data")])
def descriptive_statistics(independent_variable, dependent_variable):
    ind_variable = pd.read_json(independent_variable, orient = "split")
    dep_variable = pd.read_json(dependent_variable, orient = "split")
    ind_dep_variable = pd.concat([ind_variable, dep_variable], axis = 1)
    descritpive_statistics = ind_dep_variable.describe()
    descritpive_statistics = descritpive_statistics.reset_index()
    data = descritpive_statistics.to_dict("rows")
```

```
    columns = [{"name": i, "id": i} for i in (descritpive_statistics.columns)]
    return dt.DataTable(data = data, columns = columns, style_table =
    {"overflow": "auto",
                                                    "striped":"True",
                                                    "bordered":"True",
                                                    "hover":"True"})
```

Specifying a Callback for Correlation Analysis Results

Listing 11-19 specifies a callback for correlation analysis results. First, it loads the independent variables and the dependent variable by implementing the read_json() method. It then discerns the correlation coefficients using the Pearson correlation method and passes them into a table.

Listing 11-19. Specifying a Callback for Descriptive Statistics

```
@app.callback(Output("linear-model-predictions", "children"),
              [Input("independent-variables-values", "data")],
              [State("dependent-variable-value", "data")])
def linear_model_predictions(independent_variable, dependent_variable):
    ind_variable = pd.read_json(independent_variable, orient = "split")
    dep_variable = pd.read_json(dependent_variable, orient = "split")
    ind_dep_variables = pd.concat([ind_variable, dep_variable], axis = 1)
    x = ind_dep_variables.iloc[:,0:-1].values
    y = ind_dep_variables.iloc[::,-1].values
    y = np.array(y)
    x_train, x_test, y_train, y_test = train_test_split(x , y, test_size =
    0.2, shuffle = False)
    linear_model = LinearRegression(n_jobs = -1)
    linear_model.fit(x_train, y_train)
    linear_model_predictions = pd.DataFrame(linear_model.predict(x_test),
    columns = ["Forecast"])
    linear_model_predictions = linear_model_predictions.head(20)
    linear_model_predictions = linear_model_predictions.reset_index()
    data = linear_model_predictions.to_dict("rows")
```

```
    columns = [{"name": i, "id": i,} for i in (linear_model_predictions.
    columns)]
    return dt.DataTable(data = data, columns = columns, style_table =
    {"overflow": "auto",
                                                  "striped":"True",
                                                  "bordered":"True",
                                                  "hover":"True"})
```

Specifying a Callback for an Algorithm's Predictions

Listing 11-20 specifies a callback for an algorithm's predictions. First, it loads the independent variables and the dependent variable by implementing the `read_json()` method. Next, it preprocesses the data, trains the ordinary least-squares algorithm, generates predictions, and passes it into a table.

Listing 11-20. Specifying a Callback for an Algorithm's Predictions

```
@app.callback(Output("linear-model-predictions", "children"),
              [Input("independent-variables-values", "data")],
              [State("dependent-variable-value", "data")])
def linear_model_predictions(independent_variable, dependent_variable):
    ind_variable = pd.read_json(independent_variable, orient = "split")
    dep_variable = pd.read_json(dependent_variable, orient = "split")
    ind_dep_variables = pd.concat([ind_variable, dep_variable], axis = 1)
    x = ind_dep_variables.iloc[:,0:-1].values
    y = ind_dep_variables.iloc[::,-1].values
    y = np.array(y)
    x_train, x_test, y_train, y_test = train_test_split(x , y, test_size =
    0.2, shuffle = False)
    linear_model = LinearRegression(n_jobs = -1)
    linear_model.fit(x_train, y_train)
    linear_model_predictions = pd.DataFrame(linear_model.predict(x_test),
    columns = ["Forecast"])
    linear_model_predictions = linear_model_predictions.head(20)
    linear_model_predictions = linear_model_predictions.reset_index()
    data = linear_model_predictions.to_dict("rows")
```

```
    columns = [{"name": i, "id": i,} for i in (linear_model_predictions.
    columns)]
    return dt.DataTable(data = data, columns = columns, style_table =
    {"overflow": "auto",
                                                    "striped":"True",
                                                    "bordered":"True",
                                                    "hover":"True"})
```

Specifying a Callback for an Algorithm's Intercept and Coefficients

Listing 11-21 specifies a callback for an algorithm's intercept and coefficients. First, it loads the independent variables and the dependent variable by implementing the read_json() method. It then preprocesses the data, trains the ordinary least-squares algorithm, generates the intercept and coefficients, and then passes it into a table.

Listing 11-21. Specifying a Callback for an Algorithm's Intercept and Coefficients

```
@app.callback(Output("linear-model-intercept-and-coefficients",
"children"),
              [Input("independent-variables-values", "data")],
              [State("dependent-variable-value", "data")])
def linear_model_intercept_and_coefficients(independent_variable,
dependent_variable):
    ind_variable = pd.read_json(independent_variable, orient = "split")
    dep_variable = pd.read_json(dependent_variable, orient = "split")
    ind_dep_variables = pd.concat([ind_variable, dep_variable], axis = 1)
    x = ind_dep_variables.iloc[:,0: -1].values
    y = ind_dep_variables.iloc[::, -1].values
    y = np.array(y)
    x_train, x_test, y_train, y_test = train_test_split(x , y, test_size =
    0.2, shuffle = False)
    linear_model = LinearRegression(n_jobs = -1)
    linear_model.fit(x_train, y_train)
    linear_model_predictions = linear_model.predict(x_test)
    labels = str("Coef_") + ind_variable.columns
```

```
intercept = pd.DataFrame(pd.Series(linear_model.intercept_))
intercept.columns = ["Intercept"]
coefficients = pd.DataFrame(linear_model.coef_).transpose()
coefficients.columns = labels
intercept_coefficients = pd.concat([intercept, coefficients],axis=1).
transpose()
intercept_coefficients.columns = ["Values"]
intercept_coefficients = intercept_coefficients.reset_index()
data = intercept_coefficients.to_dict("rows")
columns = [{"name": i, "id": i,} for i in (intercept_coefficients.
columns)]
return dt.DataTable(data = data, columns = columns, style_table =
{"overflow": "auto",
                                            "striped":"True",
                                            "bordered":"True",
                                            "hover":"True"})
```

Specifying a Callback for an Algorithm's Evaluation Results

Listing 11-22 specifies a callback for an algorithm's evaluation results. First, it loads the independent variables and the dependent variable by implementing the read_json() method. After that, it preprocesses the data and trains the ordinary least-squares algorithm and generates the predictions. Finally, it constructs a table that contains results relating to the algorithm's performance.

Listing 11-22. Specifying a Callback for an Algorithm's Evaluation Results

```
@app.callback(Output("linear-model-evaluation", "children"),
              [Input("independent-variables-values", "data")],
              [State("dependent-variable-value", "data")])
def linear_model_evaluation(independent_variable, dependent_variable):
    ind_variable = pd.read_json(independent_variable, orient = "split")
    dep_variable = pd.read_json(dependent_variable, orient = "split")
    ind_dep_variables = pd.concat([ind_variable, dep_variable], axis = 1)
    date = ind_dep_variables.index
```

```
    x = ind_dep_variables.iloc[:,0: -1].values
    y = ind_dep_variables.iloc[::, -1].values
    y = np.array(y)
    x_train, x_test, y_train, y_test = train_test_split(x , y, test_size =
    0.2, shuffle = False)
    linear_model = LinearRegression(n_jobs = -1)
    linear_model.fit(x_train, y_train)
    linear_model_predictions = linear_model.predict(x_test)
    mean_absolute_error = metrics.mean_absolute_error(y_test, linear_model_
    predictions)
    mean_squared_error = metrics.mean_squared_error(y_test, linear_model_
    predictions)
    root_mean_squared_error = np.sqrt(MSE)
    R2 = metrics.r2_score(y_test, linear_model_predictions)
    explained_variance_score = metrics.explained_variance_score(y_test,
    linear_model_predictions)
    mean_gamma_deviance = metrics.mean_gamma_deviance(y_test, linear_model_
    predictions)
    mean_poisson_deviance = metrics.mean_poisson_deviance(y_test, linear_
    model_predictions)
    linear_model_evaluation = [[mean_absolute_error, mean_squared_error,
    root_mean_squared_error, R2, explained_variance_score, mean_gamma_
    deviance, mean_poisson_deviance]]
    linear_model_evaluation_data = pd.DataFrame(linear_model_evaluation,
                                                index = ["Values"],
                                                columns = ["Mean absolute
                                                error", "Mean squared error",
                                                "Root mean squared error", "R2",
                                                "Explained variance score", "Mean
                                                gamma deviance", "Mean Poisson
                                                deviance"])
   data = linear_model_evaluation_data.to_dict("rows")
   columns = [{"name": i, "id": i,} for i in (linear_model_evaluation_
   data.columns)]
```

```
return dt.DataTable(data = data, columns = columns, style_table =
{"overflow": "auto",
                                                    "striped":"True",
                                                    "bordered":"True",
                                                    "hover":"True"})
```

Running the Dash App

Listing 11-23 runs the dash app by implementing the `run_server()` method and specifying mode as "`external`", including `dev_tools_ui` and `dev_tools_props_check` as `False` so that it does not debug the app prior to running it.

Listing 11-23. Specifying a Callback Function for URL Routing

```
app.run_server(mode="external",
               dev_tools_ui = False,
               dev_tools_props_check = False)
```

Conclusions

This chapter introduced integrating a machine learning algorithm into a web app. It explained a procedure for developing an algorithm (i.e., imputation and preprocessing, which includes splitting data and scaling). Note that linear regression algorithms make strong assumptions about the data (i.e., linearity and normality). This chapter did not test key assumptions. Learn more at `www.apress.com/gp/book/9781484268698`.

CHAPTER 12

Deploying a Web App on the Cloud

This chapter presents one way to deploy a web app. First, it summarizes an integrated development environment (IDE) useful for developing, testing, and debugging Python web frameworks. Subsequently, it explains how to organize the file structure before deploying a web app and presents a practical example of improving web app deployment.

Integrated Development Environment

Most Python programmers, data scientists, and machine learning models use Jupyter Notebooks. Although the Jupyter Notebook provides many benefits, like ease-of-use, it has its shortcomings. For instance, it does not provide robust debug functionalities.

If you intend to develop, test, and deploy, I advise you to use an integrated development—a platform that enables programmers to write, assemble and code an application. There are several IDEs for Python programming (i.e., Microsoft Visual Studios, Eclipse, and PyCharm, among others. This chapter uses the PyCharm IDE. First, ensure that you have installed PyCharm on your computer. Download the installation file from the official PyCharm website. It supports Microsoft Windows, macOS, and Linux.

PyCharm

PyCharm is a prevalent Python IDE that backs HTML, JavaScript, and SQL, among others. It works similarly to other IDEs, like Microsoft Visual Studios, but JetBrains progressively developed it for virtual environment management, debugging, and assembling Python programming code. Besides that, it adequately supports distribution

T. C. Nokeri, *Web App Development and Real-Time Web Analytics with Python*,
https://doi.org/10.1007/978-1-4842-7783-6_12

packages like Anaconda, which is the most prevalent distribution package for managing Python libraries.

Download the PyCharm installer from the JetBrains official website (`www.jetbrains.com/help/pycharm/installation-guide.html`) and install it on your computer. Then, start the IDE and create a new project after installing PyCharm on your local computer. Figure 12-1 shows the File" in the PyCharm IDE.

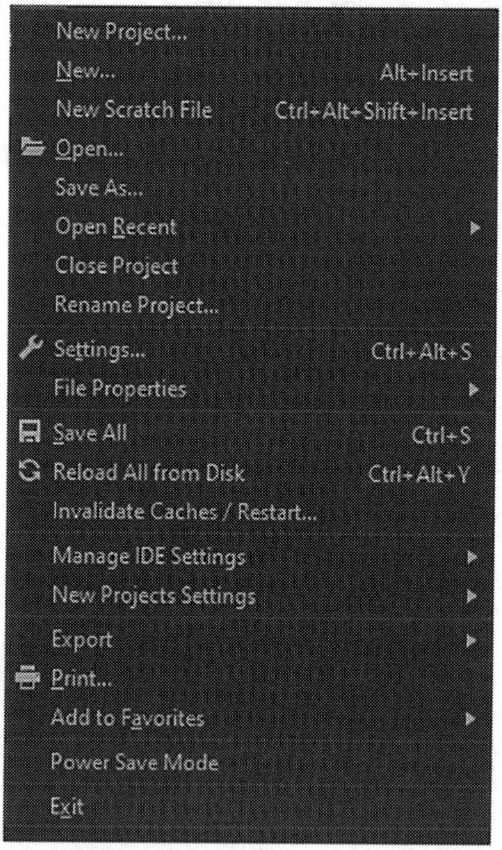

New Project...
New... Alt+Insert
New Scratch File Ctrl+Alt+Shift+Insert
Open...
Save As...
Open Recent
Close Project
Rename Project...
Settings... Ctrl+Alt+S
File Properties
Save All Ctrl+S
Reload All from Disk Ctrl+Alt+Y
Invalidate Caches / Restart...
Manage IDE Settings
New Projects Settings
Export
Print...
Add to Favorites
Power Save Mode
Exit

Figure 12-1. *Creating a new project*

Virtual Environment

When developing more complex web apps, you may need to decompose the code to separate Python files that interact with each other. A virtual environment and manages scripts and libraries. After creating a virtual environment, install key libraries. Use the `pip` instance line to install the libraries. Learn more at `www.jetbrains.com/pycharm/`.

Figure 12-2 shows the Create Project window in the PyCharm IDE.

***Figure 12-2.** Specifying the interpreter: new environment*

Figure 12-2 shows that you should specify the local directory for the project, including the base interpreter (either new or existing). You can choose to create a main.py script. Clicking the Create button results in what's shown in Figure 12-3.

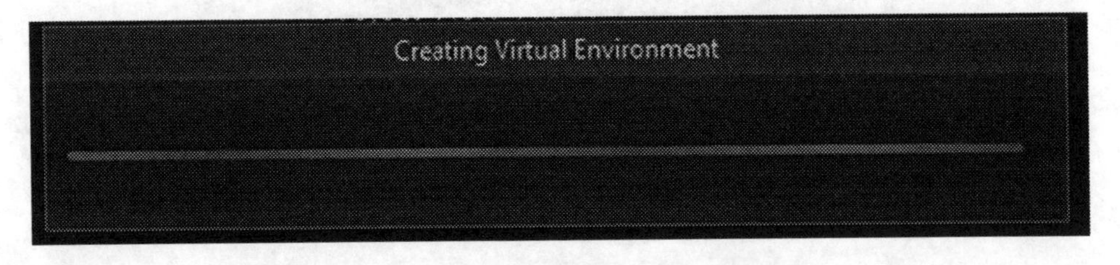

***Figure 12-3.** Creating the virtual environment creation*

Listing 12-1 initializes a Git repo, then constructs and activates a virtual environment (if you are implementing Heroku).

Listing 12-1. Construct and Activate a Virtual Environment

```
$ virtualenv venv
$ git init
$ source venv/bin/activate
```

File Structure

There is a specific way to organize folders to deploy a dash web app. Figure 12-4 is a screenshot of organized files in a Python virtual environment.

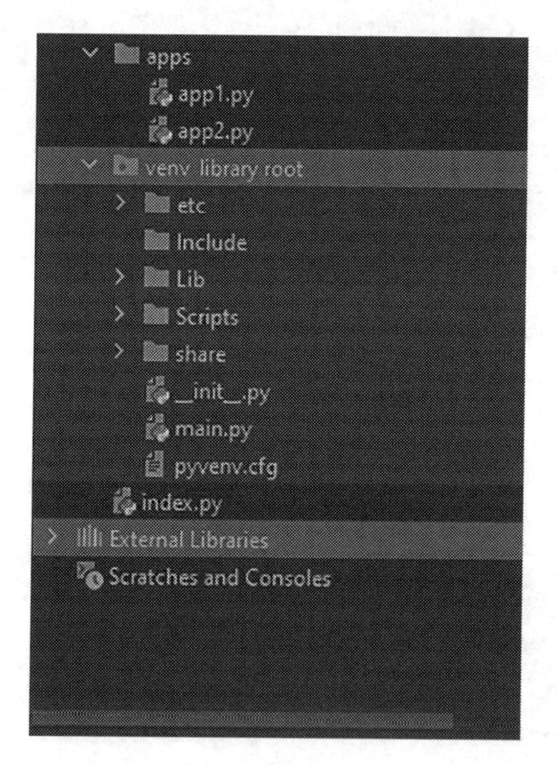

Figure 12-4. *File structure*

Figure 12-4 shows that a Python file named `__init__` contains no code.

Next, construct a Python file named `main` that contains the logical code of the main file. Then, construct an apps folder that will contain other Python files.

Integrating Innumerable Python Files

You need to separate the code into several apps when handling a large project for better code management. Each app invariably comprises a layout that routes from the main app. Listing 12-1 is an example of integrating innumerable apps.

Listing 12-2. Integration Innumerable Python Files

```
from apps import app1
```

Listing 12-2 imports `app1` from the `apps` folder (refer to Figure 12-4).

Hosting Web Apps

After developing and testing the web app, the next step involves deploying it to a specific host. For small-scale web, you host the web app on your local computer. Alternatively, you may host a web app on an online or offline web server.

Dash Enterprise

Dash has an enterprise version called Dash Enterprise, which provides services on a subscription basis. It lets you develop and deploy dash web apps on in-premise Linux servers and cloud platforms like Google Cloud, Amazon Web Services, and Microsoft Azure). It also provides a service offering related to Kubernetes, reporting, and user analytics.

Heroku

The official Heroku platform is the most prevalent free cloud platform that supports Python. It deploys web apps to a suitable Linux container. Visit the official Heroku website to register for a free account and receive 550 hours of usage. In addition, it offers services for a paid subscription compared to a free subscription.

Other cloud platforms are prevalent for app deployments, like Microsoft Azure, Amazon Web Services, and Google Cloud Platform. Heroku's key difference is that it is valuable for small-scale to medium-scale projects, while the others are convenient for medium-scale to large-scale projects.

To properly deploy an app, you initially deploy it to a popular version-control platform known as Git. Although it amply provides a value-add through its ease of use, other platforms offer web services and computational power that make scaling projects relatively easy.

Listing 12-3 constructs a folder by implementing `$ mkdir` and `$ cd` and specifies the folder name.

Listing 12-3. Construct a Folder

```
$ mkdir dash_web_app
$ cd dash_web_app
```

Listing 12-4 installs key libraries.

Listing 12-4. Install Key Libraries

```
$ pip install dash
$ pip install plotly
```

Listing 12-5 installs `gunicorn`, which in common is a dedicated WSGI HTTP server that duplicates files and arranges them for web app deployment through positioning the master and facilitating requests.

Listing 12-5. Install Gunicorn

```
$ pip install gunicorn
```

Listing 12-6 initializes the `dash_web_app` using `$ heroku`. Following that, `$ git add. dash_web_app` includes files to the local Git directory. Subsequently, `$ git commit -m` typically commits the source web app source code to the local Git directory. Afterward, `$ git push heroku` master drives the web app source code to the Heroku master (or branch). Finally, it scales the web app.

Listing 12-6. Deploy Heroku

```
$ heroku dash_web_app
$ git add.dash_web_app
$ git commit -m
$ git push heroku master
$ heroku ps:scale web=1
```

Conclusion

This concludes a book that introduced creating interactive dashboards as web apps (integrated with machine learning models). I believe that you have learned enough to take your code into production.

Index

T. C. Nokeri, *Web App Development and Real-Time Web Analytics with Python*,
https://doi.org/10.1007/978-1-4842-7783-6

M

N, O

P, Q

R

S

T

U

V

W, X, Y, Z

Printed in the United States
by Baker & Taylor Publisher Services